SCIENCE, F.
CLIMATE CRISIS

Praise for *Science, Faith and the Climate Crisis*

'In our supposedly secular age, when humans are transforming the world's physical geography like never before, we need alternative thinking about the nature of the problems that confront us. Religious and secular thinking must make space for a dialogue of equals, so that we can identify ways forward on Earth that are feasible, are desirable and possess legitimacy. This book of thoughtful essays represents, and might in turn help to build, such a dialogue.'

Professor Noel Castree, University of Manchester, UK

'A timely and important contribution to "why" and "how" science and faith need to collaborate to combat the climate crisis.'

Willy Telavi, Former Prime Minister of Tuvalu

SCIENCE, FAITH AND THE CLIMATE CRISIS

EDITED BY

SALLY MYERS
Woolf Institute, UK

SARAH HEMSTOCK
Bishop Grosseteste University, UK

EDWARD HANNA
University of Lincoln, UK

emerald PUBLISHING

United Kingdom – North America – Japan – India
Malaysia – China

Emerald Publishing Limited
Howard House, Wagon Lane, Bingley BD16 1WA, UK

First edition 2020

British Library Cataloguing in Publication Data
A catalogue record for this book is available from the British Library

ISBN: 978-1-83982-987-1 (Print)
ISBN: 978-1-83982-984-0 (Online)
ISBN: 978-1-83982-986-4 (Epub)

Printed and bound by CPI Group (UK) Ltd, Croydon, CR0 4YY

ISOQAR certified Management System, awarded to Emerald for adherence to Environmental standard ISO 14001:2004.

ISOQAR REGISTERED
Certificate Number 1985
ISO 14001

INVESTOR IN PEOPLE

CONTENTS

LIST OF FIGURES

ABOUT THE AUTHORS

The Rt Rev Dr Marc Handley Andrus is the eighth bishop of the Episcopal Diocese of California. He was installed as bishop in 2006 – a position of oversight for a diocese comprised of 24,000 communicants serving the San Francisco Bay Area of California. Prior to his election as Bishop of California, Andrus served as Bishop Suffragan in the Episcopal Diocese of Alabama. Bishop Andrus represents the Most Rev. Michael Bruce Curry, the Presiding Bishop for the Episcopal Church, at the UN Conference of Parties on Climate Change (2015 to present), and is a member of the Leaders Circle for the *We're Still In* network.

Sheila Moore Andrus, PhD, Episcopal Diocese of California Sustainability Team, USA, is an environmental scientist, science manager and educator with a focus on global health, climate change and sustainable development. She has led entomological research for the USDA Forest Service, and global health research and capacity building at the University of Alabama at Birmingham's (UAB) Sparkman Center for Global Health. Her consulting includes work with the University of California San Francisco's Global Health Program. She participated in the UN Conference of Parties on Climate Change, 2015–2018, and leads rollout for the Episcopal Church's web-based tool (sustainislandhome.org) to address climate change.

Dr Mark Charlesworth is an Associate Tutor at Bishop Grosseteste University, UK, and his research focuses on policy for rapid climate change including the theological implications.

Dr Michael Colechin is Founder of Cultivate, a company he set up to support organisations and individuals who are seeking to deliver innovative, low carbon, energy solutions. He brings a creative and dynamic approach to exploring the challenges we face in reducing our impact on the environment and preventing catastrophic climate change. His work is rooted in his experience of working as an engineer researching and developing solutions to these issues with industry, academia and government. He delivers workshops that are designed to be provocative and thought-provoking, exploring how we can all be part of the solution.

Dr Emily Colgan is Senior Lecturer in Biblical Studies at Trinity Theological College in Aotearoa, New Zealand. Her research focuses on the relationship between the Bible and contemporary social imaginaries, asking about the degree to which the ideologies contained within biblical texts continue to inform communities in the present. Emily is particularly interested in ecological representations in the Bible, as well as depictions of gender and violence.

Dr Patrina Dumaru lectures at the University of the South Pacific, Fiji, and provides technical advice to Pacific Island governments on national resilient development planning and monitoring and evaluation processes. She has 20 years of applied research and consultancy experience in the Pacific region covering coastal management, mobility, water and sanitation, environmental impact assessment and gender issues.

Fetalai Gagaeolo is a Disaster Management Officer with the Government of Samoa. She has participated in the UNFCC COP meetings supporting Pacific region delegations.

Dr Richard J. Hall is a Research Associate in the School of Geographical Sciences, University of Bristol, UK. He previously worked as a Research Associate at the University of Lincoln, UK. His interests are in North Atlantic jet stream variability, seasonal forecasting for North Western Europe and the impact of the changing Arctic upon the mid-latitudes. His recent work focuses on stratospheric variability and impacts on extreme weather.

Prof. Edward Hanna is Professor in Climate Science and Meteorology in the School of Geography, University of Lincoln, UK. He is an international authority on Greenland climate change and mass balance of its ice sheet, contributed to the Intergovernmental Panel on Climate Change's 2013 Fifth Assessment Report, and has co-authored many Arctic Report Cards of the US National Oceanographic and Atmospheric Administration. He is the World Climate Research Programme Climate and Cryosphere (CliC) representative on the ISMASS (Scientific Committee on Antarctic Research/International Arctic Science Committee/CliC Expert Group on Ice Sheet Mass Balance and Sea Level) Steering Committee. Prof. Hanna received the *International Journal of Climatology* Editor's Award from the Royal Meteorological Society in 2018, and in 2020 he was a co-recipient of the Institute of Marine Engineering, Science and Technology's Denny Medal. He co-led the 2019 'Moana: Water of Life: Navigating Climate Change for Planetary Health' conference in Lincoln.

Dr Peni Hausia Havea is a Tongan Scholar and a training Pastor from the Kingdom of Tonga in the South Pacific.

Dr Sarah Hemstock is a Programme Leader for Geography at Bishop Grosseteste University, UK and an Adjunct Fellow of the University of the South Pacific. She believes education is the best tool to achieve resilient development and has a consuming enthusiasm for her subject. For the UNFCCC, she has been a national party member and Pacific Community representative to the Paris Committee on Capacity Building. She has served on various task forces for EU and UN climate change (mitigation, adaptation and migration) and disaster reduction initiatives. She has also developed Pacific regional tertiary educational policy for resilient development; adopted by 26 countries and territories. In 2010 she was made a Government of Tuvalu Honorary Ambassador – Officer for Environmental Science.

Siu Fanga Jione has expertise in Participatory Geographic Information System (PGIS) and applications of GIS in vulnerability assessment and DRM. She has experience in training and capacity building at community and national levels, utilising participatory tools, methods and approaches.

Dr Andrew P. Kythreotis is a Senior Lecturer in Social and Political Geography in the School of Geography, University of Lincoln, UK, and a Senior Researcher at the Tyndall Centre for Climate Change Research, University of East Anglia, UK. His research and teaching revolves around the broad themes of climate change and the environment and how its policy, politics and governance is constructed around socio-spatial ontologies. Subject specialisms include Climate Change Policy, Politics and Governance; Political and Social Geography; Political Ecology of the State; Environmental Knowledge Construction; Scale Debate in Human Geography; Evidenced-based Policy and Open Science. He has advised national governments on climate change adaptation issues, has recently co-founded the Lincoln Climate Commission and was an external reviewer for DEF-RA's UK Climate Change Risk Assessment the 2017 and upcoming 2022 Evidence Reports.

John Lewin, in addition to his current visiting professorship at Lincoln, is Emeritus Professor of Physical Geography at Aberystwyth University, UK, where he also served as Dean of Science, Vice Principal and Pro-Vice-Chancellor. His main research interests are in the geomorphology, sedimentation and use histories of rivers and floodplains. A current interest lies also in student learning needs at university, particularly in light of the developing environmental changes arising from global heating.

Professor Mark G. Macklin is the Foundational Head of the School of Geography at the University of Lincoln, UK. Prof. Macklin also oversees the Lincoln Centre for Water and Planetary Health, a pioneering research centre which focuses on solving the most pressing global environmental and societal problems emerging from the world's largest rivers. His research interests include river channel and floodplain responses to climate change, long-term human–river environment interactions, alluvial archaeology, flood risk assessment, metal mining pollution and its impact on ecosystem and human health and the hydrological controls of malaria. In 2018, Prof. Macklin received the Murchison Award of the Royal Geographical Society for his pioneering research in fluvial geomorphology and its environmental applications.

Lynnaia Main serves as the Episcopal Church Representative to the United Nations. Along with other representatives, she nurtures partnerships between The Episcopal Church and UN entities, member states and civil society organisations. Since 2015, she has collaborated with Episcopal Church leaders to enhance the Church's climate and environmental action and advocacy within the UN system. This has included accompanying Episcopal delegations to annual UN Climate Change Conferences since COP21 in Paris and shepherding the Church's application for and admission in 2017 as an observer organisation to the UN Framework Convention on Climate

Change (UNFCCC). Her work and ministry on the Church's behalf cover a range of UN issues and moments touching upon water, environment, climate and sustainable development. Prior to joining the Church, Ms Main was a university lecturer in international relations and political science. The research presented here draws from these backgrounds.

Dr Theresa G. Mercer is a Senior Lecturer in Biogeography and Planetary Health at the School of Geography, University of Lincoln, UK. She has held several academic postings at Cranfield University, UK, Cardiff University, UK, Keele University, UK, the University of Queensland, Australia, and the University of Hull, UK. She is an interdisciplinary environmental scientist with broad interests in environmental management and Education for Sustainable development (ESD).

Rev Dr Sally Myers is a priest in the Church of England and an academic. She has worked with a number of HE institutions in the strategic development of teaching and research programmes. Her own research and professional interests are concerned with how individuals and communities represent their faith to themselves and others, how this impacts attitudes and behaviour, and particularly, how this changes over time in response to learning and crisis to form 'wisdom'. She draws upon narrative and cognitive psychology to understand the mechanisms of change involved in 'changing minds' and the effectiveness of educational interventions. She is currently a Visiting Scholar at the Woolf Institute, Cambridge. At the time of the 2019 conference, she was Principal of the Lincoln School of Theology.

Connor Price is reading Geography at Bishop Grosseteste University, UK. He has a passion for researching climate change and the impacts it will have on societies, especially within the Pacific region.

ACKNOWLEDGEMENTS

The authors would like to thank the organisers of the 2019 Moana: Water of Life Conference: the Diocese of Lincoln and the University of Lincoln, UK, in particular the Lincoln Centre for Water and Planetary Health, and also Bishop Grosseteste University, UK, and the Lincoln Faith and Environment Group for their significant contributions in making the event a meeting of minds and hearts. We also thank Sorina Hanna for help with figure redrafting.

FOREWORD

It becomes ever clearer that climate change is the greatest challenge that we and future generations face; a true horseman of the apocalypse. This book comes at a crucial time. Without swift, decisive action, the consequences of climate change will be devastating.

On a trip to Fiji last year, I was told by my hosts, 'For you Europeans climate change is a problem for the future; for us, it is a problem of everyday survival'. The contributions of our Polynesian friends in this book illustrate how, for many, climate change is already climate crisis. And, Polynesia is just one example of the repercussions of climate change we are seeing worldwide. In Nigeria, desertification has caused conflict amongst tribes competing for dwindling resources. In Bangladesh, monsoons have killed thousands. There are many many more examples.

As a Christian, I believe in the words of Psalm 24, 'The earth is the Lord's, and all that is in it'. We are the stewards of God's creation and it is our sacred duty to protect the natural world, which we have so generously been given. Responding to climate change is an essential part of this responsibility. But it is not only that. When we look at Jesus we see one who instinctively stood alongside the most vulnerable in society. It is absolutely clear that following Jesus today must include standing alongside those who are on the front line of this unfolding catastrophe.

The situation is difficult, but it is not hopeless. There is still a chance to act. With prayer and fruitful discussions, fresh inspiration for action that makes a lasting impact can be found.

I am constantly inspired and encouraged to hear of the passionate, creative and committed ways individuals and churches are living out their faith, working to address the causes of climate change and to reduce its effect. It is happening across our global Anglican Communion, and it will rightly be a central part of our conversations at the forthcoming Lambeth Conference.

However, although important action is already being taken at local, national and international levels, there is much more to be done, and it needs to be done far more urgently. We must continue to speak out and act. We also need to highlight the wealth of academic research on this subject and at the same time amplify the voices of those who are suffering and living with the daily impacts of climate change. I am delighted that this book is doing just that.

The book follows on from the international conference on climate change held in Lincoln in 2019. The conference was a collaboration between the University of Lincoln, UK, Bishop Grosseteste University, UK (an Anglican foundation), and the dioceses of Lincoln and Polynesia. I think that the partnership between religious institutions and science can make a profound difference in facing many of the world's problems, not least this one. The combination and collaboration of expertise, global reach and diverse experience is a real and powerful route to substantial change. This book illustrates how when different voices are listened to carefully, new perspectives, opportunities and solutions can begin to be found.

Justin Welby
Archbishop of Canterbury

PREFACE

Creation is God's gift to everyone. The climate is a global phenomenon. Whoever we are and wherever we live on this incredible planet of ours, we share a responsibility to protect the environment and to do so as people connected with one another across geographical and cultural boundaries. Living in isolation is no longer an option for any of us, and neither can we retreat into the bliss of ignorance. Scientists have been naming the issue and telling us what is happening to the environment for a long time. Responding to climate change and taking seriously our stewardship of God's creation is an essential part of our responsibility as human beings.

Some parts of the world are more exposed than others to the effects of climate change, and there is a particular threat to the low-lying islands of Oceania from rising sea levels. This book follows on from the 2019 Moana: Water of Life Conference, which brought together academics, educators and faith leaders from Aotearoa, New Zealand, and the Diocese of Polynesia, with whom the Diocese of Lincoln is linked, and the Episcopal Church of the United States. Both projects set out to be deliberately collaborative in nature and to provide an opportunity to hear first-hand from those who come from places and cultures different from our own. I hope and pray that in listening to each other we will be challenged in our thinking and encouraged in our actions and that together we will gain a deeper understanding of what God is calling the church to be and do as the Body of Christ at this time in human history.

The Rt Rev Dr David Court
Acting Bishop of Lincoln

INTRODUCTION

Earth's climate is changing faster than at any point in the history of modern civilisation through human-induced global warming, resulting in damaging weather extremes across all borders, social and geographic. Our combined decisions on greenhouse gas (GHG) usage will determine the size of the carbon footprint inherited by future generations. Yet there remains significant reluctance to respond robustly to compelling scientific evidence on human-induced climate change. When the subject is mentioned, many people tend to retreat into everyday economic, social and political issues and cultural conflicts of interest. This is at least in part a way of avoiding difficult-to-hear information from others. One way to help unblock this impasse is to bring together diverse stakeholders for intentional conversation.

The 2019 conference Moana: Water of Life: Navigating Climate Change for Planetary Health brought together physical and social scientists, educators, students, theologians and activists, including many who are already living with the effects of climate change in the Pacific Island Region (PIR) in face-to-face conversation. Following on from the conference, this edited volume brings together papers from some of the contributors to that conversation and examines the nature, challenges and potential fruits of future dialogue between these different stakeholders.

The conference was an example of Talanoa, a Fijian word, which is used across the Pacific to describe a process of round-table (or rather round-bowl) participatory, inclusive, 'blame-free', open and honest discussion. Deep receptive listening to, and reflective dialogue with, those with different perspectives is difficult and requires sustained concentration. This is further exacerbated when issues are emotive, urgent, and where there are conflicting priorities and no clear solution. The conference, and this book, set out to address this difficulty. The volume deliberately includes and takes seriously diverse perspectives to explore and elucidate complex understandings and attitudes, in this instance specifically towards water and the climate crisis. It includes chapters that are written by authors who are deeply grounded in the Christian faith and authors who do not profess a faith. It offers perspectives from very different kinds of research and academic writing in science, theology and education. Practical solutions are offered alongside spiritual reflections. Whilst all of the contributors are academics, no one discipline, methodology or epistemological foundation is taken as normative. What is privileged, however, is the voice of the young, and especially of the students involved in research projects focused on and based in the PIR.

As with the voices heard at the conference then, the chapters in this book are authentic to their authors. Beyond referencing, their contributions have not been edited to a proscribed template. Nor are they presented as a string of ideas neatly arranged and pre-digested for easy consumption, as this would suggest a cohesion that does not exist in the reality outside of the pages. Rather, in an attempt to recreate the delight and creativity of genuine dialogue the editors invited each author to write 'in their own voice'. We believe that one of the most important ways of honouring another human being is to truly listen to what they have to say, for it is in the messiness of real encounter that real understanding might begin to emerge. The book comprises 12 chapters and invites the reader to 'listen' in turn to perspectives from students living and researching on the front line of the climate crisis as it unfolds in the PIR, climate scientists, climate educators, climate theologians and people offering practical solutions to the climate crisis. The final chapter then offers an overview and exploration of the contributions. It sets out the challenges involved in dialogue between stakeholders with diverse backgrounds and world views, drawing out common themes and assessing areas of divergence, and provides an interpretative overview of emerging patterns together with suggestions for taking the conversations forward.

The first two chapters are written by young researchers of deep faith from the PIR. In Chapter 1, Fetalai Gagaeolo explores the socio-cultural impacts of relocation due to climate change with reference to the people of Kiribati moving to Fiji. Local community perceptions and experiences reveal that relocation is considered the last 'adaptation resort', which will compromise their socio-cultural practices and values in the long term. Communities have spiritually come to terms with the climate crisis and just do not want to relocate.

In Chapter 2, Peni Hausia Havia considers the effect of climate change on PIR livelihoods, health, overall well-being and sustainability through the lens of a deep Christian faith in God. He argues that no current development model addresses the relationship between physical 'climate change impacting factors' and God. He outlines how his research led to the creation of a framework called 'Moana: Nothing is impossible with God' (Luke 18:27) for use by community leaders and policy-makers in tackling the PIR response to the climate crisis.

The next two chapters offer two complementary inputs by leading environmental scientists from the Lincoln Centre for Water and Planetary Health, University of Lincoln, UK. In

Chapter 3, Mark Macklin and John Lewin set the scene with an overview of how rivers have played a defining role in the global development of human societies and culture. They explore how this will undoubtedly continue in the twenty-first century with a growing demand for water, increasing pollution of river channel and floodplain environments, and anthropogenic global warming-related changes in the frequency of floods and droughts. They consider how rivers initially shaped societies, and then how urbanisation, industrialisation and intensified agriculture have more recently transformed river systems, compromising planetary health and human ways of life. They highlight likely areas facing the greatest challenges and offer insight into how interdisciplinary catchment-based approaches, and new technologies such as those based on satellite imagery, are now beginning to address pressing societal and planetary problems in the unfolding climate crisis.

In Chapter 4, Edward Hanna and Richard Hall recount the overwhelming scientific evidence that human activity through enhanced GHG emissions is largely responsible for recent climate change and accompanying extreme weather. They discuss the scientific evidence for recent climate change, focusing on the very rapid environmental changes recently seen in the Arctic and Antarctic icy realms that affect global sea level rise and may also influence extreme weather events in densely populated Northern Hemisphere mid-latitude regions. They then consider further likely changes in climate during the rest of this century. In the final section, they briefly propose necessary actions and solutions by the global community including religious stakeholders, arguing that fully involving religious communities is essential in order to help decarbonise the economy.

The next two chapters offer two different approaches to climate change education. In Chapter 5, Sarah Hemstock,

Siu Jione, Mark Charlesworth and Patrina Dumaru argue that in order to help society at large adapt to and survive the climate crisis, it is important to foster climate change education. They explore the international policy context which has led to ground-breaking research and Pacific educational initiatives (such as recognising climate change and disaster risk reduction as employment sectors, leading to the development of regional vocational qualifications, and regional accreditation of qualifications) before considering a Pacific Small Island Developing States (P-SIDS) case study.

In Chapter 6, Theresa Mercer and Andrew Kythreotis discuss how society can be more involved in climate research and policy by fostering a more socially equitable and just way of tackling future climate impacts through education. They outline how previous and contemporary social and political conditions relate to increased and fairer citizen engagement in climate action in the science-policy domain. They then explore how collaborative education approaches through Education for Sustainable Development (ESD) might increase citizen engagement in climate action. Their chapter concludes by critically discussing future directions for research in ESD and climate change for a more inclusive and just form of climate governance.

The following two chapters are written by two very different theologians. In Chapter 7, Marc Andrus, Bishop of the Episcopal Diocese of California, argues that the water crisis is in fact a spiritual crisis. He draws together personal experience and scripture, especially the story of the flood in Genesis and the account of Jesus walking on the water in Mark's Gospel, to reflect on the relationship between humanity and the rest of creation. He details some of the devastating consequences of our continued disregard of the earth's delicate ecosystem, including freshwater scarcity, plastic pollution, species loss and endangerment and climate refugees. He ends with a review of the role of faith bodies in addressing the water crisis, through theological interpretation and practical action.

Chapter 8 is written by biblical scholar Emily Colgan of Trinity Theological College in Auckland, New Zealand. She outlines the concept of an ecological interpretation of scripture. She argues that one way to cut through the all-too-human tendency to ignore or answer away the need for personal and corporate climate action is to connect directly with the religious beliefs held by people as a result of the interpretation of scripture. Her chapter offers a compelling way of thinking about coherence between scripture and religion and their relationship with the natural environment. She argues for a creation-centric reading of the Bible and in doing so suggests that the preservation of the natural world and reversal of climate change is a realistic and indeed urgent response to the authority of scripture.

The following three chapters offer three very different practical responses to the climate crisis. In Chapter 9, Mike Colechin explores the role that technology may have in addressing global warming. He argues that a range of solutions will be required, involving changes to social practices and the development of energy supply solutions that have a net zero impact on GHG emissions. He then explores a range of potential changes that individuals can make to their daily lives in support of the wider transformation needed within society as a whole.

In Chapter 10, Sheila Andrus observes that although the climate crisis is frightening for many people, it is often not clear what one person, or one community, can do to drive down GHG emissions. She introduces a practical solution in the form of an Internet accessible tool, sustainislandhome.org, that can help people and communities be part of climate solutions. Her chapter focuses on why sustainislandhome.org was developed, its design principles and how it works, and the lessons the Episcopal Church is learning from the rollout of this tool across Episcopal dioceses in the United States.

In Chapter 11, Lynnaia Main looks at the role played by faith-based organisations in discussions at the United Nations. She offers a brief history of faith in action at the UN, including a comprehensive overview of the history and structures involved. She then outlines success stories, future challenges and potential obstacles. Finally, she explores how people of faith can work more effectively with and within the United Nations to address climate change.

In Chapter 12, Sally Myers looks back over the first 11 chapters of the book and offers an analysis of the different voices and perspectives they contain. She briefly sets out the challenges involved in bringing together and deeply listening to those with diverse backgrounds, priorities and traditions. She then explores the main themes that have emerged from the chapters and conversations, offering an overview of areas of difference, but also of surprising fundamental agreement, not just on the 'what', but also the 'how' of what needs to happen next.

The conclusion draws together the dialogue between all those who participated in the conference. It makes suggestions as to how people of faith, and the Anglican Church in particular, might respond to the challenge of climate change in 'thought, word and deed'. It then identifies areas for further investigation, dialogue, decision and action.

There were two significant contributions to the 2019 conference that have not yet been mentioned but are listed below:

The Diocese of Lincoln launched its Environmental Policy at the 2019 conference, which may be found at https://www.lincoln.anglican.org/Handlers/Download.ashx?IDMF=277ea34d-9e7b-480a-8d75-5250dd47a121.

Project Pressure provided an exhibition visualising the climate crisis: https://www.project-pressure.org.

1

CHRISTIANITY, KIRIBATI AND CLIMATE CHANGE-INDUCED MIGRATION

Fetalai Gagaeolo, Sarah Hemstock, and Connor Price

INTRODUCTION

The impacts of climate change on societies and cultures within the Pacific are extensive, far reaching and rapid (Hausia Havea, Hemstock, & Jacot Des Combes, 2018; IPCC, 2018). There are a range of factors that contribute to increasing vulnerability of small island developing states in the Pacific Islands region to climate change, including geographic isolation, ecological uniqueness and fragility, human population pressures and associated waste disposal problems. It is important to note the centrality of Christianity to Pacific culture and the likely implications this holds for understanding and addressing climate change (Capstick, Hemstock, & Senikula, 2018; Hausia Havea, Hemstock, Jacot Des Combes, & Luetz, 2017).

A survey of 1,226 students at the University of the South Pacific, Fiji, found that over 80% of them attended church at least on a weekly basis; this is a sample of the region's educated urban elite, and therefore its future climate leaders (Nunn, 2017). This factor coupled with the climate crisis the region faces makes the relationship of religion and climate change an essential facet to research. The argument for a closer consideration of the role of religion in tackling climate change in the Pacific Islands has begun to be been taken up by researchers, who have suggested that religion has an essential role to play in how people make sense of climate change in the region (Capstick et al., 2018; Hausia Havea, Hemstock, Jacot Des Combes, Luetz, & Liava'a, 2019; McLeod, 2010). Therefore, this chapter looks to explore this relationship with a focus on the spiritual dimension of climate change-induced migration (from Kiribati to Fiji) and its associated cultural impacts.

BACKGROUND

It is inevitable that 'migration' would be viewed as an adaptation strategy and Kiribati was the first nation to prepare for climate-induced migration. In 2014, the Church of England sold their 20 km^2 land holding in Fiji to the Government of Kiribati, who bought it in a flurry of press attention and political gesture with the intention of using it firstly to improve Kiribati's food security, and secondly as a place to relocate to.

In July 2017, in his opening address at the United Nations Framework Convention on Climate Change, COP23 President, the Prime Minister of Fiji Frank Bainimarama, in an awesome speech, stated:

> *...No one living in the Pacific can be left in any doubt about what is at stake. ... As Pacific Islanders, we are fighting for our very survival. ...*

> *Of all the vulnerable nations of the world, you [Kiribati and Tuvalu] are the most vulnerable....*
>
> *And even if the battle to keep your islands above the water is lost, we will continue to stand shoulder- to-shoulder with you. Fiji has offered to give permanent refuge to the people of Kiribati and Tuvalu, our closest neighbours under threat. ....*
>
> *In a worst-case scenario, we know that there will be climate refugees throughout the world. But the people of Kiribati and Tuvalu will not be refugees. We Fijians will embrace you and take you into our homes and our hearts. That is our solemn promise to you as Pacific neighbours and friends. ...*
>
> (Bainimarama, 2017)

Following on from this momentous commitment from Fiji, in 2018, New Zealand legally recognised the status of 'climate refugees' and offered 100 visas a year to Pacific Islanders on that basis (Office of the Minister of Foreign Affairs, 2019). It is a token gesture since it would take more than a 1,000 years to evacuate Kiribati's current population of approximately 116,000 people, but the recognition of 'climate refugee' status is important. Oakes, Milan, and Campbell (2016) found that in Kiribati, 94% of households had been impacted by natural hazards between 2005 and 2015, with 81% being affected by sea level rise. Relocation seems a sensible option in the face of those impacts.

However, this study finds that Kiribati communities (I-Kiribati) are not willing to leave. It is apparent that many Pacific Islanders are ready to accept the impacts of climate change on their islands and are opposed to forced relocation which they recognise as compromising their solid foundation and faith in God.

> *I was born as a Kiribati, so I will die as a Kiribati. I don't care if I would die poor here on my own land. Just wait, if it's God's will to wash Kiribati away, so be it. We do not know the hour and the day for the end of time to arrive. We can relate this to Noah's time. We have to blame ourselves because of the rumours of climate change that have been ongoing; it is a sign that we have changed a lot. Climate is changing because humans are changing; we are modifying the natural resources from God and living rebelliously.*
>
> (#15 Male, Abaiang, Kiribai)

It would be a mistake to overlook these views given their prevalence among the communities in Kiribati, where more than 90% of the population identify as Christian, and the Pacific as a whole (Nunn, 2017). These views on climate change are complex and to some extent can be demonstrated through multiple formulations of the Noah story with corresponding ethical and political imaginaries (Fair, 2018). However, these accounts provide locally meaningful and morally compelling counter-narratives of Pacific regional climate change issues.

STUDY DESIGN

The design of this study was non-experimental. Most of the participants involved in this research were randomly selected with a few selected key informants irrespective of age, sex and status. A total of 56 interviews were conducted. Primary data were qualitative and record personal experiences and perceptions of climate and environmental changes. Relocation was explored in relation to how it would influence socio-cultural issues with specific reference to Fiji as the host country. The study was conducted in Kiribati where core information and data were collected from two islands: South Tarawa and Abaiang.

EMERGING THEMES

The study sets out to discuss the spiritual dimension of relocation experienced by I-Kiribati moving from Kiribati to Fiji.

From interviews it emerged that spirituality enabled people to cope with climate change impacts in Kiribati and could potentially provide successful adaptation strategies. Most studies on climate change-induced mobility (Australian Government, 2014; Donner, 2015; Remling, 2020) focus on migration patterns or the political, environmental, socio-economic and scientific aspects. Previous studies have little emphasis on the concerns and justifications of those who refuse to migrate and specifically, no studies were found which examine those who refuse to migrate on a spiritual basis. It is important to note that the refusal, based on spiritual beliefs, of many I-Kiribati to relocate to Fiji does not mean that they are in denial or climate sceptic. Their refusal is rooted in their awareness of dangerous climate change impacts worldwide, so efforts to escape it are viewed as pointless. In response to this acceptance, study participants were actively adapting to climate change 'spiritually'; hence building up their spiritual persona by 'realigning their ways to God's decrees' which they believe to be their best adaptation option. Peoples' justifications are based on their own Christian biblical interpretations which they used to explain their real-life experiences.

Responses indicated three themes to explain how people are coping with climate change spiritually. First, people have strong 'faith in God' to defend them in this climate crisis;

second is their strong belief that as long as their 'inner being' is secure and in tune with God, then the direct threats of climate change on their physical being are not a worry; and third was the biblical belief that climate change was one of the apocalyptic signs of the world's end times. Since the majority of the responses were from people with Christian backgrounds, they formulated conclusions based on the Bible to justify their views. Based on their faith, the Bible provides the normative and definitive basis for interpretation of all events. For those that adhere to such interpretations, it reveals the secrets to the unseen fixed and predetermined future that awaits the world, including climate change.

Theme 1 – A Strong Connection to and 'Faith in God' as the Source of Refuge

The participants who refused to relocate were very optimistic about their decisions despite their awareness of their vulnerability to climate change. One of the responses was,

> *Wait, because at the moment there is still not yet a clear sign that Kiribati would submerge. Relocation can be the answer right now, but we still have to wait on what the Lord has for Kiribati.*
>
> (#12 Male, South Tarawa)

> *For us, we believe that Jesus is coming soon. If we're faithful to Him, then the problem won't matter, because we also have another 'home in heaven'. We have a lot of plans like building a ship but if we have faith then I'm sure God will see us through, we have to be prepared.*
>
> (#21 Male, South Tarawa)

Given their interpretation of God as the creator of nature, it is understandable why the grassroots are claiming God to be more trustworthy in delivering them from climate change impacts than human scientific knowledge and physical solutions. One respondent mentioned,

> *We run here and there to our neighbours but they are also affected. So that was the chance for us to look up to God for a heavenly solution. Maybe God has allowed climate to change to draw people closer to Him.*
>
> (#10 Female, South Tarawa)

Theme 2 – 'As Long as My Inner Nature Is Saved and in Tune with God's Will'

The second justification was the belief that

> *...as long as the inner nature or spiritual person is saved and in tune with God's will, then the means of relocating for the security and safety of the physical nature is not obligatory.*

One of the detailed responses was,

> *No (I won't move), it is better I stay and die here in Kiribati rather than dying in a foreign country that we have no knowledge about. Fiji cannot save me from dying; we will all die no matter where we run so it's better I die here. I will not move as long as my inside person is safe and secure no matter what would happen with my physical being.*
>
> (#5 Female, South Tarawa)

Another respondent mentioned,

> *Wait, I will not go as long as I am saved in the inside. Kiribati will not be submerged because God says in His bible that there would be no more flooding.*
>
> (#35 Male, South Tarawa)

These responses created more questions than they answered, namely, 'What does it mean, "as long as the spiritual person is saved", and what is the "spiritual person" and what are they being saved from?' Their understanding appeared to be that no physical life would be spared in the wrath of climate change, and that running away from it cannot reverse its consequences. This study revealed that the security of their spiritual nature has given insights into a legitimate aspect of human security in the face of climate change.

> *We believe that Jesus is preparing the best home for us in heaven, so I think it's best to prepare the person inside or the spiritual person so that when the end times arrive or you die, your inside person goes straight to heaven where there is joy, peace and love, no more crying, sorrow or even death.*
>
> (#5 Female, South Tarawa)

Another respondent also mentioned, […]

> *'If we're faithful to Him (God), then the problem won't be that much because we have another home in heaven'.*
>
> (#33 Male, South Tarawa)

Possibilities were also expressed with regard to biblical references to a destructive future based on the participant's spiritual views.

> *Physical is physical, saving our lives physically such as relocating to other places would not save us from the end of the world that is at hand. Now is the time to be prepared because the kingdom of heaven is near.*
>
> (#5 Female, South Tarawa)

Such compelling views reveal that local communities of Kiribati are already looking at climate change beyond science and politics, and furthermore, view human security from an eternal point of view. There were many biblical sources referred to as evidence for these perceptions – Amos 5: 19; Zephaniah 1:2-3; Zephaniah 1: 14-16.

Theme 3 – Anthropogenic Climate Change: An Apocalyptic Sign of the End times and an Appointed Time for All to 'Return to God'

There seemed to be a linkage between climate change and the end times according to the responses. Although climate change is anthropogenic, through the lens of Kiribati people, it is allowed by God as a warning for all humanity to repent and return to Him and that the end times and the establishment of His Kingdom is at hand. Hence in Kiribati, climate change is not just an indication of a physical crisis but primarily a spiritual crisis. One respondent mentioned,

> *I don't want to move, God placed me here, I will not lose thinking about my own land but whenever I move I would always think as a Kiribati until the very last day; these are the signs of the end times that Jesus is coming soon, and we have to get ready and repent.*
>
> (#10 Female, South Tarawa)

Another respondent also mentioned, quoting from the Bible,

> *Yes, in the book of Matthew 24, it mentions all the signs of the end times and of Jesus' coming, for example pestilences and so forth. Climate change is also a sign of the end times.*
>
> (#21 Male, South Tarawa)

Another also responded,

> *Climate change is very much related also to the signs of the end times we are in. At times we talk about this problem, it is one of the signs that the end of times is coming.*
>
> (#12 Male, South Tarawa)

People's reasoning based on biblical interpretation stresses that the world is in a new era of changes.

> *We have to blame ourselves because of the rumours of climate change been ongoing; it is a sign that we have changed a lot. Climate is changing because humans are changing; we are modifying the natural resources from God and living rebelliously.*
>
> (#15 Male, Abaiang)

Another interview was based on Matthew 24,

> *...the people during Jesus time asked, What is the sign of the end of the world, then Jesus replied, the stones that you see, there will be no stone on top of another, in other words, the stones will be washed off similarly to what we see now with our sea walls because of sea level rise. So this is one of the signs that the world should be aware of, that the end of time is drawing near.*
>
> (#21 Male, South Tarawa)

CONCLUSIONS

In exploring relocation it was brought to light that climate change, through the lens of Kiribati communities, is not just a matter of physical, environmental, political or economic crisis but a spiritual crisis. Kiribati people are looking beyond the physical challenges and interpreting climate change from a spiritual dimension in order to understand and take action accordingly. Through their strong spiritual connections and beliefs, I-Kiribati have confidence and faith to remain on their islands rather than relocate. Paradoxically, spirituality could hold the key to successful relocation. It is, therefore, essential to include considerations of religion and spirituality and the role of churches in the settlement of Pacific migrants, as also argued by McLeod (2010). Religion and spirituality are the foundation of Kiribati communities and the spiritual side of the 'self' is a significant feature of climate change adaptation.

With international dialogues around climate change branching into many diverse issues, Kiribati people have considered human security from an eternal viewpoint, which does not prevent practical interventions for successful adaptation if relocation becomes necessary. Rather it defines the essential values that will make an adaptation strategy successful. Therefore, the views and justifications of those who refuse to migrate or relocate from a spiritual angle should not be ignored in climate change-induced mobility or anthropology studies.

REFERENCES

Australian Government. (2014). Recognized seasonal employer workers arrival into New Zealand: Year 2013/2014. Retrieved from http://employment.govt.nz/er/rse/information/arrival-2013-2014.pdf

Bainimarama, F. (2017, July 3). Opening address at the Climate Action Pacific Partnership event, UNFCCC COP23. Retrieved from https://unfccc.int/news/pm-bainimarama-opening-address-at-the-climate-action-pacific-partnership-event

Capstick, S., Hemstock, S., & Senikula, R. (2018). Perspectives of artist–practitioners on the communication of climate change in the Pacific. *International Journal of Climate Change Strategies and Management*, *10*, 323–339.

Donner, S. D. (2015). The legacy of migration in response to climate stress: Learning from the Gilbertese resettlement in the Solomon Islands. *Natural Resources Forum*, *39*(3–4), 191–201.

Fair, H. (2018). *'Not drowning but fighting': Faith, activism, and climate change narratives in the Pacific Islands.* Doctoral dissertation, UCL (University College London).

Hausia Havea, P., Hemstock, S. L., & Jacot Des Combes, H. (2018). Improving health and well-being through climate change adaptation. In L. T. Duncan (Ed.), *Advances in Health and Disease. Nova Biomedical* (Vol. 4, pp. 215–230). Hauppauge, NY: Nova Science Publishers Inc. ISBN:987-1-53613-346-2

Hausia Havea, P., Hemstock, S. L., Jacot Des Combes, H., & Luetz, J. (2017). "God and Tonga are my inheritance!": Climate change impact on perceived spirituality, adaptation and lessons learnt from Kanokupolu, 'Ahau, Tukutonga, Popua and Manuka in Tongatapu, Tonga. In W. Leal Filho (Ed.), *Climate change impacts and adaptation strategies for coastal communities* (pp. 167–186). Climate Change Management – Springer Nature Series. Cham: Springer International Publishing. ISBN:978-3-319-70702-0

Hausia Havea, P., Hemstock, S. L., Jacot Des Combes, H., Luetz, J., & Liava'a, L. F. C. (2019). Impact of climate change on spirituality. Macmillan Brown Centre for Pacific Studies. *Pacific Dynamics: Journal of Interdisciplinary Research*, *3*(1), 33–45. ISSN:2463-641X

IPCC. (2018). In V. Masson-Delmotte, P. Zhai, H.-O. Pörtner, D. Roberts, J. Skea, P. R. Shukla, … T. Waterfield (Eds.), *Special report: Global warming of 1.5°C. An IPCC special report on the impacts of global warming of 1.5°C above pre-industrial levels and related global greenhouse gas emission pathways, in the context of strengthening the global response to the threat of climate change, sustainable development, and efforts to eradicate poverty.* Geneva: Intergovernmental Panel on Climate Change. Retrieved from https://www.ipcc.ch/sr15/download/#chapter

McLeod, D. (2010). Potential impacts of climate change migration on Pacific families living in New Zealand. In B. Burson (Ed.), *Climate change and migration south Pacific perspectives* (pp. 135–157). Wellington: Institute of Policy Studies, Victoria University of Wellington.

Nunn, P. (2017). Side-lining god: Why secular climate projects in the Pacific Islands are failing. *The Conversation*. Retrieved from https://theconversation.com/sidelining-god-why-secular-climate-projects-in-the-pacific-islands-are-failing-77623

Oakes, R., Milan, A., & Campbell, J. (2016). *Kiribati: Climate change and migration – Relationships between household vulnerability, human mobility and climate change*. Report No. 20. United Nations University Institute for Environment and Human Security (UNU-EHS), Bonn.

Office of the Minister of Foreign Affairs. (2019). Cabinet environment, energy and climate committee. Pacific climate change-related displacement and migration: A New Zealand action plan. Retrieved from https://www.mfat.govt.nz/assets/Uploads/Redacted-Cabinet-Paper-Pacific-climate-migration-2-May-2018.pdf

Remling, E. (2020). Migration as climate adaptation? Exploring discourses amongst development actors in the Pacific Island region. *Regional Environmental Change*, *20*, Article No. 3. doi:10.1007/s10113-020-01583-z

2

MOANA: 'OKU MAFEIA HE 'OTUA 'A E ME'A KOTOA PE (MOANA: NOTHING IS IMPOSSIBLE WITH GOD (LUKE 18:27)): REFLECTIONS OF A TONGAN EARLY CAREER RESEARCHER ON GOD AND CLIMATE CHANGE

Peni Hausia Havea

INTRODUCTION

As a scholar from Tonga, I developed this reflection based on my own experiences as a training Pastor for the Free Wesleyan Church (FWC), working in Pacific Island region on the area of climate change and religion. I am taking a quote from the Bible (Luke 18:27) to use as a theme to tell the story of Pacific people about Moana and how this state-of-the-art idea helps in shaping Pacific Island Resilience (Climate Change

Adaptation and Disaster Risk Reduction – CCA & DRR). In the context of the Pacific (e.g. Tonga), the meaning of Moana is complex. From my experience, 'Moana' can be expressed in numerous forms: (1) can be attributed to the ocean or the open sea; (2) can be utilised as a name (e.g. a person, canoe) and (3) metaphorically, it could mean warrior, ideas, collaboration, oneness or partnership (Miyose, 2019; Streiff & Dundes, 2017). Using this idea as a guide, I have developed the 'Moana: Nothing is impossible with God' framework to guide the implementation of Sustainable Development Goal (SDG) and Resilience (CCA & DRR) initiatives in the Pacific. If government and civil society agencies were to implement this model while implementing community activities, Resilience Moana and SDGs could be achieved by 2030.

As an early career researcher interested in the perceived and physical impacts of climate change in Pacific Island communities and their adaptation, I welcome this opportunity to reflect on my personal journey and influences in my own voice. My ebullience in this domain emerged from my PhD which studied perceived impacts of climate change on livelihoods, health and well-being (e.g. physical, mental and spiritual) of 460 Tongans aged 15 to 75 from five coastal communities, Kanokupolu, 'Ahau, Tukutonga, Popua and Manuka, in the main island, Tongatapu. On this journey, I discovered that because Tonga is a Christian State, people's faith (e.g. denominations) influences their perceptions of Climate Change Adaptation (CCA). As a result, their paradigm of Moana inherits similar interpretations, whether from the Bible or teachings from their denominations (Havea, 2020).

In this reflection, as is the way in the Pacific, all the meanings of Moana will be used interchangeably, with the reader encouraged to interpret the expression of 'Moana' wherever necessary – basing their interpretation on the passion they feel for Moana. Now, because Moana can also comport a persona of a human soul or spirit (e.g. protector), from the perspective of climate change, this reflection views Moana as a living thing too.

The role of religion in Pacific societies is pervasive and can be overwhelming – this is the norm. Most Pacific Islands nations are Christian States; even in the secularised and multi-faith states (e.g. Fiji), the Christian faith has considerable influence on every aspect of society and life in general. Religions also play a significant role in politics (Cox, Finau, Kant, Tarai, & Titifanue, 2018; Nunn et al., 2016). For me personally, coming from a family where God is always ab initio and a nation where God is a priority, faith has influenced me socially, economically, politically and academically (Havea, Hemstock, Jacot Des Combes, & Luetz, 2018; Michaelis, 2011). For us in the Pacific, this influence is totally acceptable. This kind of Moana, however, may not be acceptable in other societies (e.g. Australia, the United Kingdom) (Morrison, Duncan, & Parton, 2015; Nunn et al., 2016) and is understandable – we all have our differences and do our best to accept them.

From what has been presented in the literature (Havea et al., 2018; Kuruppu, 2009), no studies have identified the linkages between climate change, Moana and God. However, for Pacific Islanders like myself, these concepts are at the forefront of dealing with climate change impacts and improving planetary health generally.

More importantly, virtually, no model has been developed to realise an interplay between the perceived role of 'God' (Christian and/or non-Christian) in 'Resilience' (CCA & DRR). We could ask 'what is the perceived motivations – saving your soul, saving your house and land, a future for your children?' Linking peoples' motivations with positive actions could hold the key to meeting the 2030 Development Agenda (UNDP, 2015) as well as provide a sustainable and resilient Moana by 2030. Linking people's perceptions and motivations is at the centre of my research.

Luke 18:27 states that 'The things are impossible with people are possible with God'. This statement demonstrates how faith in God is perceived as meeting the adaptation needs of Pacific Islanders. For instance, a woman from Manuka in Tonga reflecting on climate change, Moana and God, said: 'Our family priority is going to church. God controls everything in this universe'. Another woman from Kanokupolu took the discussion deeper in associating religion with science:

> *...I do believe these climate change and impacts are part of Gods' plan. Because [there is] nothing that God cannot do – '...ko e 'Otua 'oku ne mafeia 'a e me'a kotoa pe...'. What we need to do though as Christians is to follow exactly what the scriptures dictated us to do. Nature is a true representation of God himself. If we try to solve nature [i.e., climate change] by our human intelligence [i.e., science], we can't do it. Like Tonga, we should be proud that our great King George Tupou and I have devoted that 'my God and Tonga are my inheritance'. So, worship and prayer are always part of our culture, so we need God to help us understand nature and solve this problem.*
>
> (Havea, 2020, p. 178)

Based on these excerpts, people use their faith to personally adapt to the effects of climate change by turning to God for comfort in the understanding that science and the State does not always provide them with all they needed – as opposed to their God. This pairing of religion and science is always missing from the assessment of climate change and Moana by the scientific community, although for Pacific islanders, the role of faith is vital (IPCC, 2014a). The scientific community needs to understand there is also a spiritual part of one's life, and as a result, this part will always be affected by climate change. As demonstrated in the 'Pikipiki Hama kae Vaevae Manava' ideology, religion and science should always come together to share ideas (Havea et al., 2018; Havea, Hemstock, Jacot Des Combes, Luetz, & Liava'a, 2019; Michaelis, 2011; Nunn et al., 2016).

In responding to this important call, this reflection aims to discuss the Pacific perspectives of the role of God in improving resilience, and also, in no short order, to develop a 'nothing is impossible with God framework' for Moana!

In responding to this important call, this reflection aims to discuss the Pacific perspectives of the role of God in climate change, and also, in no short order, to develop a 'nothing is impossible with God framework' for Moana!

METHODS

For my research, a mixed method, namely Concurrent Convergence Parallel Triangulation Design (CCPTD) was used to carry out data collection, analysis, and also to model this framework (Havea, Hemstock, & Jacot Des Combes, 2017).

RESULTS AND DISCUSSIONS

Linkages between Climate Change Impacts and Moana

To comprehend how this indigenous framework works, let's start from the beginning. My PhD research examined climate change impacts on health, livelihoods and well-being. For the purposes of this reflection, some impacting factors and links have been summarised in Fig. 2.1. But before scrutinising the impacting factors, let's think again about the meaning of Moana. It may be that the linkages between climate change and Moana anticipate effective forms of adaptation and motivation/behavioural change. If you agree with me on this,

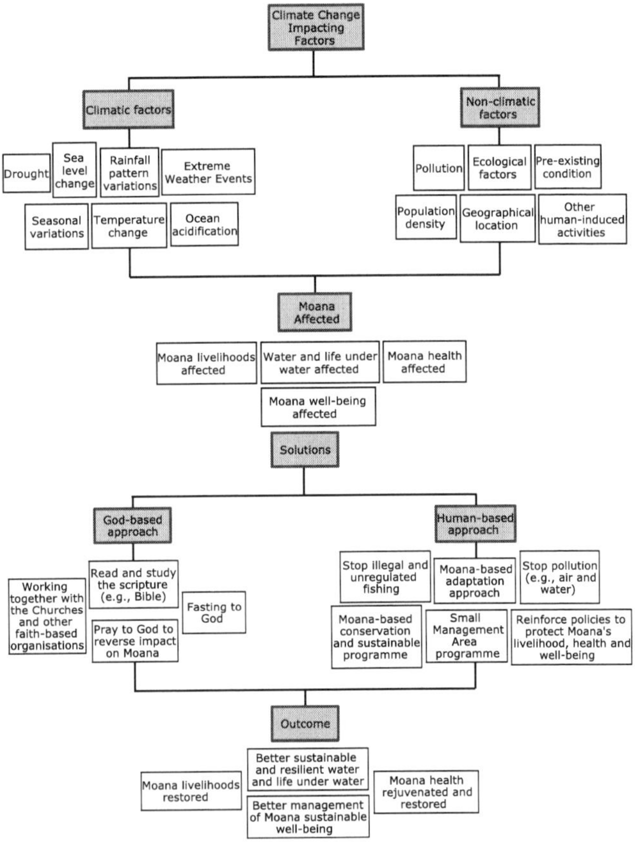

Fig. 2.1. The Linkages between Climate Change, Moana and God, Moana and Adaptation.

then the establishment of this relationship is vital to climate change and us all.

Climatic factors include natural drivers or what people may refer to as God-based factors (e.g. natural hazards) (Capstick, Pidgeon, Corner, Spence, & Pearson, 2016; IPCC, 2014b; Rohe, Schlüter, & Ferse, 2018).

Understanding how to deal with these natural factors and how they are affecting Moana is a basis for building a more resilient and sustainable Moana. One view of adaptation, it is what Charles Darwin (Darwin, 1859) has called the 'survival of the fittest'. Meaning, in the 'theory of evolution' and climate change, 'It is not the strongest of the species that survives, nor the most intelligent that survives. It is the one that is most adaptable to change' (Darwin, 1859; Hemstock & Capstick, 2019; Raoof, 2017). So, theoretically, when applying this epistemological framework to Moana as individuals, Moana can survive these climate changes if it adapts to the changing physical, social, political, moral and spiritual environment in which it finds itself.

Linkages between Moana, God and Climate Change Adaptation

Once impacting factors have been identified, they need associated solutions with desired outcomes for Moana. Most solutions relate to community-based adaptation approaches (God-based and human-based) (Fig. 2.1).

Combining the God-based adaptation approach with the following human-based adaptation capability is recommended:

- the implementation of Moana-based adaptation approaches (e.g. reforestation of mangrove);

- stopping of illegal and unregulated fishing and fishery activities;
- stopping or lessening all polluting;
- the implementation of Moana-based conservation and sustainable programme (e.g. prohibit illegal sand mining);
- implementing Small Management Area (SMA) programmes in affected communities and
- the reinforcement of relevant policies to protect Moana's livelihoods, health and well-being.

In practice, combining God-based and human-based approaches will provide the motivation for affected communities, State, Churches and Moana to unite in harmony and work collectively.

By using this approach, Resilience Moana can also be achieved by education – through church schools and community groups – and including formal education (Havea, Siga, et al., 2019; Hemstock et al., 2016; Hemstock et al., 2018).

'Moana: Nothing Is Impossible with God' Framework to Guide the Implementation of SDGs and Resilience Moana by 2030 and beyond

Combining linkages between climate change, God, Moana and adaptation (Fig. 2.1) produce a 'Moana: Nothing is impossible with God' framework (Fig. 2.2). In summary:

- identifying climate change impacting factors affecting Moana;
- identifying problems affecting Moana;
- defining solutions to be implemented;

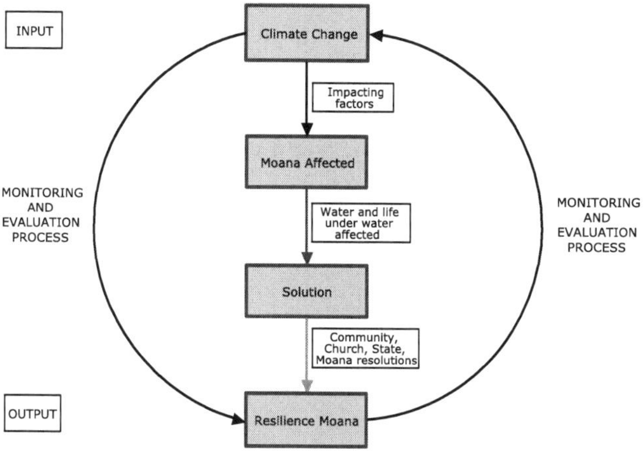

Fig. 2.2. 'Moana: Nothing Is Impossible with God' Framework to Guide the Implementation of Resilience Moana. Resilience Moana is achieved when communities are not vulnerable to climate change or natural hazards.

- defining expected solution outputs for the Resilience Moana (Fig. 2.2);
- providing motivation for behavioural change.

Additionally, if the expected outcome for the Resilience Moana is not achieved as per plan, the framework also Monitors and Evaluates (M&E) so can feedback on what went wrong, allowing improvement of processes to achieve desired outcomes.

To achieve Resilience Moana, faith and science need to work together in unison.

Achieving Resilience Moana

There are four reasons why people of the affected communities should implement this framework to achieve Resilience Moana.

First, this indigenous framework is very flexible, universal and open for interpretation. For instance, here, the model allows people to bring in their motivating factors, whether in the form of praying or using SMA to managing water and life underwater. Therefore, there is no limitation. People are the only limits by their own imaginations.

Second, this framework was constructed in favour of achieving SDG and Resilience Moana. As a result, if implemented, SDGs and Resilience Moana.

Third, the framework allows all stakeholders to participate in the building of Resilience Moana. For instance, this model permits the State, Churches, affected communities and relevant stakeholders to come together and work to achieve Resilience Moana. Therefore, any limitations of working alone will be outweighed or eliminated by uniting as one (Lata & Nunn, 2012).

Lastly, because there is power in using Jesus, this model wants to intervene by using that power to achieve Resilience Moana.

CONCLUSION

As a scholar and training Pastor for the FWC in Tonga, I realised that the Bible is a living word, whether we believe it or not. Achieving Resilience Moana is up to each individual. 'Nothing is impossible with God' is a general concept and applies to Resilience Moana and more widely. If God can tell the Camel to go through the eye of the needle and cure the blinds and lepers, this reflection believes that the same God could ensure that Resilience Moana is achievable for the universe to witness, and that's why I'm helping my community to implement adaptation strategies.

REFERENCES

Capstick, S. B., Pidgeon, N. F., Corner, A. J., Spence, E. M., & Pearson, P. N. (2016). Public understanding in Great Britain of ocean acidification. *Nature Climate Change*, *6*(8), 763–767.

Cox, J., Finau, G., Kant, R., Tarai, J., & Titifanue, J. (2018). Disaster, divine judgment, and original sin: Christian interpretations of tropical cyclone Winston and climate change in Fiji. *The Contemporary Pacific*, *30*(2), 380–410.

Darwin, C. (1859). *The origin of species*. London: John Murray.

Havea, P. H. (2020). *Climate change impact on livelihood, health and well-being: A mixed method approach*. Suva: The Unversity of the South Pacific.

Havea, P. H., Hemstock, S. L., & Jacot Des Combes, H. (2017). Preparing for better livelihoods, health and well-being – A key to climate change adaptation. In W. Leal Filho (Ed.), *Climate change adaptation in Pacific countries: Fostering resilience and improving the quality of life* (pp. 87–99). Cham: Springer International Publishing.

Havea, P. H., Hemstock, S. L., Jacot Des Combes, H., & Luetz, J. (2018). "God and Tonga are my inheritance!"—Climate change impact on perceived spirituality, adaptation and lessons learned from Kanokupolu, 'Ahau, Tukutonga, Popua and Manuka in Tongatapu, Tonga. In W. Leal Filho (Ed.), *Climate change impacts and adaptation strategies for coastal communities*. Climate Change Management. Cham: Springer International Publishing.

Havea, P. H., Hemstock, S. L., Jacot Des Combes, H., Luetz, J., & Liava'a, L. F. C. (2019a). Impact of climate change on spirituality. *Pacific Dynamics – Journal of Interdisciplinary Research*, *3*(1), 33–45.

Havea, P. H., Siga, A., Rabuatoka, T., Tagivetaua Tamani, A., Devi, P., Senikula, R., … Jacot Des Combes, H. (2019b). Using vocational education to support development solutions in the Pacific: An emphasis on climate change and health. *Applied Environmental Education and Communication*, 171–186. doi:10.1080/1533015X.2019.1569571

Hemstock, S. L., Buliruarua, L., Chan, E., Chan, G., Jacot Des Combes, H., Davey, P., … Murray, V. (2016). Accredited qualifications for capacity development in disaster risk reduction and climate change adaptation. *Australasian Journal of Disaster and Trauma Studies*, *20*(1), 15–34.

Hemstock, S. L., & Capstick, S. (2019). Communicating climate change: Reactions to adapt and survive exhibition and visitors' thoughts about climate change in the Pacific Islands region. In W. Leal Filho, B. Lackner, & H. McGhie (Eds.), *Addressing the challenges in communicating climate change across various audiences* (pp. 599–615). Cham: Springer International Publishing.

Hemstock, S. L., Jacot Des Combes, H., Buliruarua, L.-A., Maitava, K., Senikula, R., Smith, R., & Martin, T. (2018). Professionalising the 'resilience' sector in the Pacific Islands region: Formal education for capacity building. In S. Klepp & L. Chavez-Rodriguez (Eds.), *A critical approach to climate change adaptation: Discourses, policies and practices*. Advances in Climate Change Research. Routledge. ISBN 978-1-13-805629-9

IPCC. (2014a). *Climate change 2014: Impacts, adaptation, and vulnerability. Part A: Global and sectoral aspects.* Contribution of Working Group II to the Fifth Assessment Report of the Intergovernmental Panel on Climate Change. Cambridge: Cambridge University Press.

IPCC. (2014b). Topic 2: Future Climate Changes, Risk and Impacts. In R. K. Pachauri & L. A. Meyer (Eds.), *Climate change 2014: Synthesis report.* Contribution of Working Groups I, II and III to the Fifth Assessment Report of the Intergovernmental PANEL on Climate Change. Geneva: IPCC.

Kuruppu, N. (2009). Adapting water resources to climate change in Kiribati: The importance of cultural values and meanings. *Environmental Science & Policy, 12*(7), 799–809.

Lata, S., & Nunn, P. (2012). Misperceptions of climate-change risk as barriers to climate-change adaptation: A case study from the Rewa Delta, Fiji. *Climatic Change, 110*(1–2), 169–186.

Michaelis, L. (2011). Climate change and spirituality. In L. Bouckaert & L. Zsolnai (Eds.), *Handbook of spirituality and business* (pp. 236–244). London: Palgrave Macmillan.

Miyose, C. Y. (2019). "We know the way": Culture–nature relationship and Kuleana in Disney's Moana. *Popular Culture Review, 30*(2), 157–181.

Morrison, M., Duncan, R., & Parton, K. (2015). Religion does matter for climate change attitudes and behavior. *PLoS One, 10*(8), e0134868.

Nunn, P. D., Mulgrew, K., Scott-Parker, B., Hine, D. W., Marks, A. D. G., Mahar, D., & Maebuta, J. (2016). Spirituality and attitudes towards nature in the Pacific

Islands: Insights for enabling climate-change adaptation. *Climatic Change*, *136*(3), 477–493.

Raoof, A. (2017). "It is not the strongest species that survive, nor the most intelligent, but the most responsive to change": Charles Darwin. *International Journal of Anatomical Variations*, *10*(1), 0012–12.

Rohe, J., Schlüter, A., & Ferse, S. C. A. (2018). A gender lens on women's harvesting activities and interactions with local marine governance in a South Pacific fishing community. *Maritime Studies*, *17*(2), 155–162.

Streiff, M., & Dundes, L. (2017). From shapeshifter to Lava Monster: Gender stereotypes in Disney's Moana. *Social Sciences*, *6*(3), 91.

UNDP (United Nations Development Programme). (2015). *Sustainable Development Goals (SDGs)*. New York, NY: UNDP.

3

THE RIVERS OF HUMANKIND

Mark G. Macklin and John Lewin

INTRODUCTION

Water is the lifeblood of planetary health, and rivers, as a critical source of freshwater, underpin past and present human societies. River systems constantly change in response to human pressures, and protecting river channels, floodplains and their catchments from anthropogenic climate change and other human activities (e.g. pollution and dam construction), including understanding interactions with ecosystem and human health, has become a critical concern worldwide. We consider how rivers initially shaped societies, and then how urbanisation, industrialisation and intensified agriculture have more recently transformed river systems, so compromising planetary health and human lifeways. We highlight likely areas facing the greatest challenges and how interdisciplinary catchment-based approaches, and new technologies, are now beginning to address pressing societal and planetary problems in the unfolding climate crisis.

THE RIVERS OF CIVILISATIONS

The transformation of human societies from hunter-gatherers to farmers, most often associated with the creation of permanent settlements and modification of local river environments for floodwater farming, occurred in numerous parts of the world and at different times. The earliest large-scale floodplain irrigation took place in southern Mesopotamia (present-day central and southern Iraq) at c. 6000 BCE with the 'urban revolution' and creation of the first cities at c. 4000 BCE (Fig. 3.1). The other major early 'hydraulic' civilisations of the Old World in the Nile and Indus Valleys, as well as in the

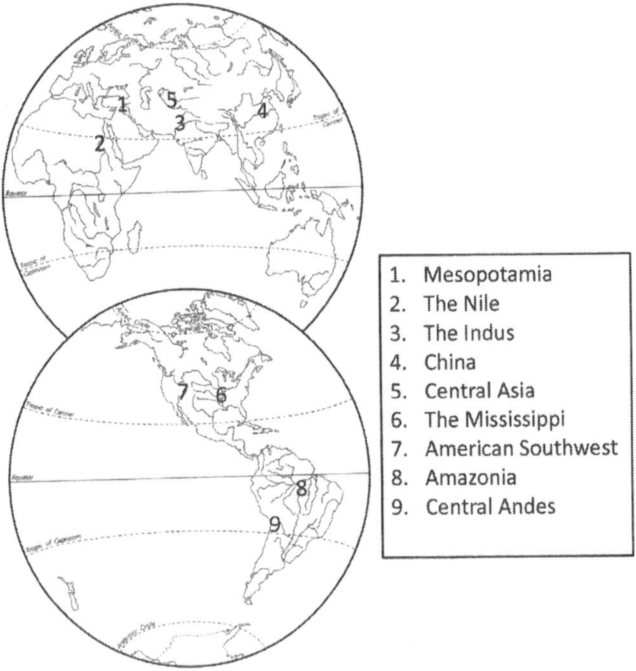

Fig. 3.1. World Rivers and Centres of Ancient Civilisations.

Huang He (China), with sophisticated water and waste management systems, socio-political structures, multiplicity of craft skills and farming surpluses, began later. Dynastic Egypt emerged at c. 3100 BCE, the Mature Harappan or fully urban phase of the Indus Valley Civilisation started at c. 2500 BCE and Dynastic China started at c.1900 BCE. Long histories of river management are also found in the Central Andean region of South America. Floodplain and sunken field agriculture was practised on the northern Peruvian coastal strip as early as c. 3000 BCE with the first stone-built true urban settlements constructed at c. 2600 BCE, followed by the introduction of large-scale irrigation at c. 1800 BCE By c. 800 BCE, all of the great mid-latitude and sub-tropical rivers of North America (Missouri–Mississippi, Colorado and Rio Grande) and Eurasia (Amu-Darya, Brahmaputra–Jamuna, Danube, Ganga, Huang He, Indus, Syr-Darya, Tigris–Euphrates, Yangtze) as well as many of those in tropical Africa (Niger and Nile), South America (Solimões-Amazon) and Southeast Asia (Irrawaddy, Mekong) supported complex agricultural societies, which increasingly indirectly and directly modified local river channels and floodplains for farming and very often for flow control and flood protection.

The development of many of these complex societies or civilisations (the word civilisation has its origins in the Latin *civis*, a citizen, itself derived from an earlier form meaning 'to settle') first took place along rivers in the arid lands of both the Americas and Asia. Choosing to 'settle down' and farm in areas of low and unreliable local rainfall may seem counter-intuitive from a modern perspective, but many of these rivers had predictable flow controlled by monsoon rains (e.g. Nile and Indus Valleys) or snow melt (e.g. Tigris and Euphrates, Mesopotamia; Amu-Darya and Syr-Darya rivers, Central Asia; Gila and Salt rivers, American Southwest; Central Andean region of South America) from mountain headwaters.

Early and pre-industrial floodwater farmers also had a high level of what might be termed 'river literacy' gained from a lifetime of watching and remembering the effects of floods and droughts on channel and floodplain dynamics, especially patterns of water flow, erosion and sedimentation that had a direct impact on the success or failure of their crops and livestock. Highly successful agricultural societies with seasonal floodplain farms and sophisticated fisheries also developed in the great river basins of the Mississippi and Missouri as well as along the Amazon River and its southern tributaries, most notably the Madeira and Xingu rivers. In the pre-Columbian period (before 1492 CE), eastern North America constituted the most spatially extensive area of pre-historic floodplain farming of any mid-latitude region, with the rivers of Amazonia supporting the largest aquaculture and floodplain horticulture system in the world.

Climate change did, however, negatively affect pre-industrial river civilisations worldwide in a similar way and sometimes more catastrophically than we have yet seen as a result of anthropogenic global warming. For instance, at c. 2200 BCE, and then again at c. 1200 BCE, as a consequence of a significant weakening of the monsoon in Africa and Asia, coinciding with extreme drought in the headwaters of the Tigris–Euphrates, the economies of river civilisations in the Huang He, Indus, Mesopotamia and Nile collapsed or went into extended decline lasting several centuries. These periods of exceptional climate volatility were also marked by very significant political and wider societal transformations. These included the fall of the Akkadian Empire at c. 2200 BCE, the breakdown of Egypt's Old Kingdom at c. 2150 BCE, the end of the urban phase of the Indus Valley Civilisation c. 2000 BCE and synchronous collapses of major Neolithic cultures in both the lower Yellow River and middle-lower Yangtze River at c. 2000 BCE The later 1200 BCE climate event coincided with the well-documented Late Bronze Age societal collapse, which was a major turning point in the history of the wider Mediterranean region, Nile Valley (end of both New Kingdom of Egypt and Kerma Kingdom of Nubia) and Mesopotamia. In the context of anthropogenic global warming and the current climate crisis, it is of concern that these episodes of hemispheric-scale societal disruption caused by both catastrophic drought and flooding were related to changes of mean global temperature of no more than c. 0.5 °C occurring over a period of more than 500 years. This compares to the 1 °C rise in mean global temperature, which has occurred since the beginning of the twentieth century in just over a hundred years.

LAND MANAGEMENT: BUILDING THE EARTH

Global perspectives on the complexities of environmental change impacts associated with past and present human activity are needed for the food and water security challenges of the twenty-first century, including those resulting from water-related diseases (Smith, Macklin, & Thomas, 2013). This is especially true for rivers, for which the onset and persistence of a range in human activities, altering their function and form, have been temporally and spatially variable. Fig. 3.2 presents an eight-category historical global assessment of land occupation as mapped by Leszek Starkel (1987), based on the pioneering research of the eminent Russian botanist Nikolai Vavilov in the 1930s (Vavilov, 1951), and updated with recent research by ourselves (Macklin & Lewin, 2019). Land cover and cultivation practice are major factors in river sediment supply (Starkel, 1987); this in turn is a major determinant of channel pattern and floodplain sedimentation of rivers (Ashworth & Lewin, 2012;

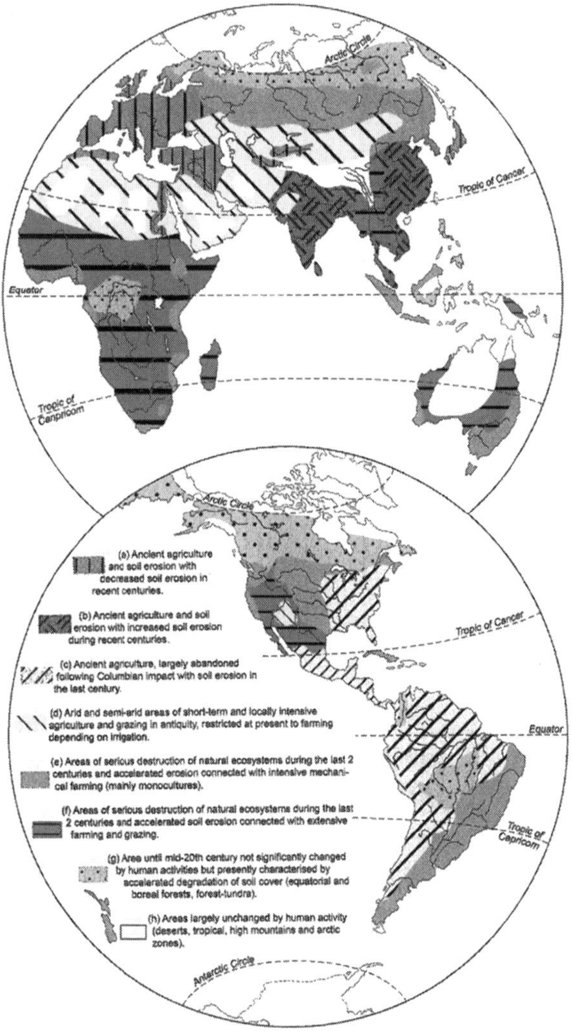

Source: From Macklin and Lewin (2019).

Fig. 3.2. Global Land-occupation Categories after Vavilov (1951) and Starkel (1987) with Additions from Macklin and Lewin (2019). Map showing areas of ancient agriculture (some abandoned and then developed more recently), those developed in the last two centuries and ones that have been little affected by human activities until the mid-twentieth century.

Macklin & Lewin, 1997). However, technologies of clearance, agriculture and management have varied considerably to produce sediment yield pulses of varying magnitudes (Ellis et al., 2013; Macklin, Lewin, & Jones, 2014). Rapid medieval population increase in Europe accelerated soil erosion, with more intensive cultivation using mould-board ploughing, but this phase was terminated by the Black Death. The Americas had landscapes 'humanised', in the sense of ecosystem change arising from anthropogenic activities, long before Columbian (post-fifteenth century) times brought in change of quite a different order (Deneven, 1992; Mann, 2005). In the tropical Americas, deforestation also succeeded a pre-European period of biomass destruction by fire followed by forest regeneration, so that 'deforestation' has not followed a simple trajectory (Nevle & Bird, 2008). Contemporary forest clearance in Amazonia now contrasts considerably in pattern, pace and technology with that of earlier eras elsewhere.

As Phillips (2001) has argued, such historical and spatial contingency means that simple prediction of human impacts is not usually possible other than in strictly situational terms. For example, future channel bank erosion responses to changing flood regimes will manifest themselves in different ways (Lewin & Macklin, 2010). Some floodplains now have long-standing deep blankets of post-settlement cohesive eroded soil materials (Macklin et al., 2014); others have been deprived of the resistance given by deep woody root systems (Gurnell, 2013). This means that riverbank resistances to flooding may now vary site by site according to what has been done to riparian vegetation and to bank cohesion.

In system terms, responses have also been pulsed and lagged as well as ramped (Phillips, 2001). For example, European prehistoric, human-induced, alluvial sedimentation lagged behind initial forest clearance until more intensive activity and new plough technology came into operation (Broothaerts,

Notebaert, & Verstraeten, 2014). For the United Kingdom, Macklin et al. (2014) collated evidence of dated sediments that possessed evidence of anthropogenic content to demonstrate a lag between the development of agriculture in the Neolithic and accelerated river sedimentation. These examples show that the effects of land use change on rivers are complex and mediated by inputs and outputs from sediment storages on hill slopes and floodplains over a millennial timescale. This is of considerable importance for the understanding of contemporary and future rivers that receive and transmit material activated by past land management – whether initially from Neolithic deforestation, medieval population increase and land pressure, colonial change or the 'Great Acceleration' of recent decades.

WHAT WE HAVE DONE TO RIVERS: WHERE AND WHEN

Although the major early civilisations – in Mesopotamia and in what are now Egypt, India and Pakistan, and China – did not start on major rivers, it was expansion along their banks and floodplains that allowed them to flourish at least from around 6,000 years ago (Macklin & Lewin, 2015). These locations commonly needed regionally organised management to provide acceptable security for settlement and extensive agriculture, and thus began a multi-millennial history of environmental transformation that has allowed human prosperity and population growth, though at some hazard to people and, of course, to partially transformed environments and their ecosystems. We live on a much-used planet (Ellis et al., 2013; Gibling, 2019).

What we have done has affected three components of river systems (Fig. 3.3). Whole land surfaces of river water

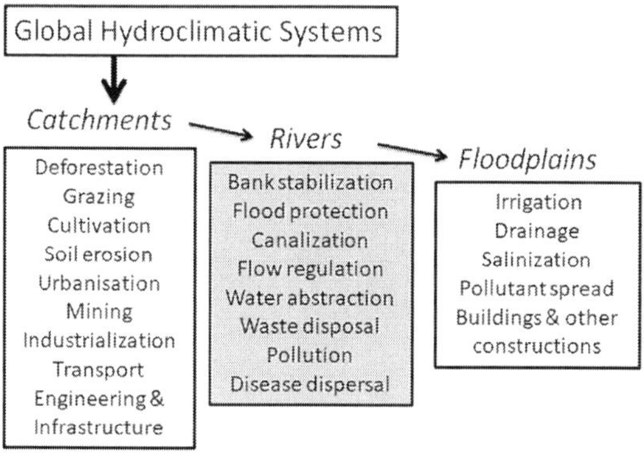

Fig. 3.3. The Activities of Humankind that Transform River Systems.

contributing areas (their catchments, or in the United States, 'watersheds') have been transformed into crop, pasture and range land, in some cases following prior clearance of forest by felling and fire. Many early environments, dependent on animal and plant domestication and grain cultivation over 7,000 years ago (Gibling, 2019, Table 2), were in former grasslands with dry climates, but even humid forests began to be cleared by 4,000 years ago (Jenny et al., 2019). This has led to accelerated erosion of exposed soil, together with all the constructed requirements of settled life: houses, workshops, streets and water supply for villages and cities, and then also the various forms of developed communication by land and water that connect them and their trade. Industrial cultures have additionally mined raw materials, extracting minerals but also spreading unwanted waste. Ancient cities have literally arisen to form 'tells' as the pile of their discarded buildings and waste has grown. Sun-dried and baked bricks provided building materials for settlements that became elevated on top of the compressed residues of earlier generations.

Secondly, river courses have been transformed in different ways: bank protection to prevent channel movement; embankments to minimise flooding; canalising and dredging to improve fleet transport for trade, but also warfare; impounding and diversion for water power and milling; and the extraction of water for consumption, irrigation and industrial processing. This has involved the re-arrangement of channels (adding new ones and eliminating others) together with weirs and reservoir development. At the same time, rivers have been used for waste conveyance leading to pollution and so to the decline of freshwater habitats and food sources, and to the spread of diseases. Large rivers, as involved with civilisations, are all different (Ashworth & Lewin, 2012), and this has led to different forms of transformation in their different centres of civilisation. For example, the Nile, with a monsoon climate regime, provided a fairly regular summer flood peak, and this allowed a controlled pattern of riparian irrigation, including devices for lifting river water up and across levees into cultivated fields. Mesopotamia, with a snowmelt regime from contributing mountains that was much less predictable in timing and amount, had a river system that was prone to avulsion into alternative courses together with a wide expanse of wetlands and lakes across the wide depression of its tectonic basin. To be fully exploited, this required intrusive management. Controlling sluices in riverside levees seems to have been a major engineering technique. The Huang He in China (aptly named as the Yellow River because of its sediment load) drained from highly water-erodible terrains of wind-deposited loess; rain also fell directly on the vast plains of North China downstream, and the need seen here was for bank protection to avoid excessive river flooding and avulsion from the sediment-depositing river. The Indus civilisations were different again: many settlements used groundwater sources along formerly more active rivers and floodplains draining from the Himalayas, with the main Indus itself being laterally unstable (Ashworth & Lewin, 2012; Giosan et al., 2012).

Thirdly, it was such floodplains that were of most significant and direct use to settled populations, so they too were transformed – draining marshes and providing irrigation channels alongside settlements. There is evidence of irrigation in c. 9400 BCE both at Jericho in the Jordan valley and on alluvial fans in what is now the mountain-margin area between Iraq and Iran. In time this has had consequences. For example, floodplain lakes in ancient Mesopotamia were once both common and rich in wildlife as food sources; these have been almost entirely infilled, whilst large areas have become salinised and useless for agriculture, both as a side effect of sediment distribution and from long-term evaporation of distributed saline waters.

These three domains have been differentially affected across the globe, with early centres of civilisations as mentioned above, but also others, such as New Zealand which was only transformed much later in the last millennium (Macklin & Lewin, 2019). Amazonia, now subject to rapid deforestation, also appears to have had early civilisations that until recently lay unknown beneath its tree canopy. The timing as well as the nature of human transformations makes a difference. For example, modern forest clearance and the laser levelling of field irrigation are very different from earlier forest felling and clearance agriculture. And the time-ordering of developments is important and reflects the technology of the times. Mechanised earth movement was largely possible only from the nineteenth century, whilst earlier human labour forces were responsible in many places for flood protection dating back many centuries. This was before the industrial revolution and a surge in manufacturing and domestic pollution that might otherwise have spread right across whole valley floors.

RIVERS OF HUMANKIND AND THE FIRST INDUSTRIAL NATION

The global variety in topography, geology, climate and settlement history has fostered much complexity in the composition of the Earth's surface, as we now see it most dramatically on satellite imagery and increasingly using unmanned aerial vehicles. An extended history also means that this surface – and in particular its river systems as discussed here – has been used by humanity in different ways. What we now see is a palimpsest of forms resulting from sequential initiatives and climate-driven episodes (Macklin, Lewin, & Woodward, 2012). Fig. 3.4 illustrates the timeline for river transformation in Britain from Roman times until the present day, accompanied by the steep rise in population in the last several hundred years. Past river modifications have

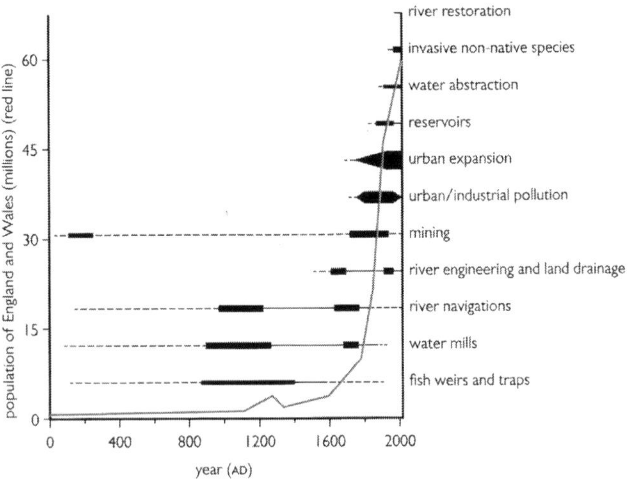

Source: From Holmes and Raven (2014).

Fig. 3.4. Timelines for Major Impacts on British Rivers over the Last 2,000 Years.

occurred as a result of the different ways in which rivers have been viewed and were generally unlike present perspectives that focus on aesthetic and recreational benefits, on ecosystem quality preservation or restoration, and on the threat of flood and drought.

Before the Norman conquest of England (1066 CE), for example, Fenland communities operated a largely self-sufficient agricultural economy based on drier 'islands' or margins set next to marsh and fen; later externally promoted drainage, in the seventeenth century and after, developed a wider market-oriented system on drained land that has become amongst the most highly valued in the country (Ash, 2017; Oosthuizen, 2017). More widely, medieval rivers performed a triple function as sources of power, as routes and facilitators of trade and as suppliers of food. Prior to the Black Death (1348–1350 CE) there was a large increase in population and land pressure, with accompanying accelerated soil erosion from land cultivated to the limit using contemporary technology. Domesday (1085 CE) also recorded over 6,000 mills, whilst fisheries, though unevenly recorded from county to county, were common (Oosthuizen 2017, Fig. 6.7). This often involved grievous conflict between mill owners and downstream farmers, as later highlighted in Magna Carta (1215 CE). On into the eighteenth century and as the economy developed, river navigations were improved and upland river tributaries became a prime source of industrial mill power before the days of steam and coal. During the late eighteenth century and first half of the nineteenth century, Britain became one of the world's leading base-metal (copper, lead and zinc) producers. One inadvertent and very environmentally damaging consequence was the large-scale contamination of UK rivers and floodplains (Lewin & Macklin, 1987). This continues to negatively affect ecosystem and human health today as the result of metal-contaminated floodplain sediments being remobilised during large flood events, which are becoming more frequent as a consequence of anthropogenic climate change (Foulds et al., 2014). Finally, urban and industrial expansion in the nineteenth century was often sited on floodplains adjacent to major cities, with accompanying water extraction and reservoir development providing safe drinking water.

This history is globally distinctive, and different world locations have their own stories. But the major point to absorb is that rivers have been used differently according to local river regimes, climate stresses, time and technology – whether supporting indigenous agricultures; more adaptive land drainage; irrigation and channel modifications; industrial development; the continental and inter-continental trade of colonial eras; or now for water supply, hydroelectric power generation and conservation for more demanding but also more ecosystem-conscious urban societies.

CONCLUSIONS

It can be argued that a realistic agenda for the world's future should neither rest on restoration of a mythic 'natural' past nor on an environmentally insouciant pursuit of ideology-based growth in nearly everything. What this chapter has outlined is an actual world as it has witnessed the millennial rise and fall of civilisations. They have evolved and adapted, and at times have suffered devastation. It is pragmatic adaptation in the light of climate and river events and episodes that has been the key to longevity. This has included environmental modification in order to maintain the living needs of growing populations, including food and water supplies, but also to cope with endemic natural hazards that have occurred under each of their fluctuating and sub-continental geophysical and biological systems. It has also required the preservation of those natural elements that are needed for the flourishing of life forms, not only but including humans. As Hanna and Hall show in the next chapter, global heating is now enhancing worldly changes generally, but not the same ones everywhere. Ice melt and thermal expansion of oceans are leading to inundation of coastlines; increasing dangers arise from floods in some areas but droughts and firestorms in others, while elsewhere there are the troubling dimensions of devastating typhoons and hurricanes. It may or may not be possible to limit global climate change, but the civilisation–river histories here outlined surely suggest that new ways of local innovative adaptation to developing change and risk must be found for community living to flourish, as we have shown happened in the past, admittedly with uneven success.

REFERENCES

Ash, E. H. (2017). *The draining of the fens*. Baltimore, MD: Johns Hopkins University Press.

Ashworth, P. J., & Lewin, J. (2012). How do big rivers come to be different? *Earth-Science Reviews*, *114*, 84–107.

Broothaerts, N., Notebaert, B., & Verstraeten, G. (2014). Non-uniform and diachronous Holocene floodplain evolution: A case study from the Dilje catchment, Belgium. *Journal of Quaternary Science*, *29*, 351–360.

Deneven, W. M. (1992). The pristine myth: The landscape of the Americas in 1492. *Annals of the Association of American Geographers*, *82*, 369–385.

Ellis, E. C., Kaplan, J. O., Fuller, D. Q., Vavrus, S., Goldewijk, K. K., & Verburg, P. H. (2013). Used planet: A global history. *Proceedings of the National Academy of Sciences*, *110*, 7978–7985.

Foulds, S. A., Brewer, P. A., Macklin, M. G., Haresign, W., Betson, R. E., & Rassner, S. M. E. (2014). Flood-related contamination in catchments affected by historical metal mining: An unexpected and emerging hazard of climate change. *The Science of the Total Environment*, *476*, 165–180.

Gibling, M. R. (2019). River systems and the Anthropocene: A late Pleistocene and Holocene timeline for human influence. *Quaternary*, *21*, 1–36.

Giosan, L., Clift, P. D., Macklin, M. G., Fuller, D. Q., Constantinescu, S., Durcan, J. A., … Syvitski, J. P. M. (2012). Fluvial landscapes of the Harappan civilization. *Proceedings of the National Academy of Sciences*, *109*, 1688–1694.

Gurnell, A. (2013). Plants as river system engineers. *Earth Surface Processes and Landforms*, *39*, 4–25.

Holmes, N., & Raven, P. (2014). *Rivers*. Oxford: British Wildlife Publishing.

Jenny, J.-P., Koirala, S., Gregory-Eaves, I., Francus, P., Niemann, C., Ahrens, B., … Carvalhais, N. (2019). Human and climate global-scale imprint on sediment transfer during the Holocene. *Proceedings of the National Academy of Sciences*, *116*, 22972–22976.

Lewin, J., & Macklin, M. G. (1987). Metal mining and floodplain sedimentation in Britain. In V. R. Gardiner (Ed.), *International geomorphology, 1986* (pp. 1009–1027). Chichester: Wiley.

Lewin, J., & Macklin, M. G. (2010). Floodplain catastrophes in the UK Holocene: Messages for managing climatic change. *Hydrological Processes*, *24*, 2900–2911.

Macklin, M. G., & Lewin, J. (1997). Channel, floodplain and drainage basin response to environmental change. In C. R. Thorne, R. D. Hey, & M. D. Newson (Eds.), *Applied fluvial geomorphology for river engineering and management* (pp. 15–45). Chichester: John Wiley & Sons.

Macklin, M. G., & Lewin, J. (2015). The rivers of civilization. *Quaternary Science Reviews*, *114*, 228–244.

Macklin, M. G., & Lewin, J. (2019). River stresses in anthropogenic times: Large-scale global patterns and extended environmental timelines. *Progress in Physical Geography*, *43*, 3–23.

Macklin, M. G., Lewin, J., & Jones, A. F. (2014). Anthropogenic alluvium: An evidence-based meta-analysis for the UK Holocene. *The Anthropocene*, *6*, 26–38.

Macklin, M. G., Lewin, J., & Woodward, J. C. (2012). The fluvial record of climatic change. *Philosophical Transactions of the Royal Society A*, *370*, 2143–2172.

Mann, C. C. (2005). *1491: New revelations of the Americas before Columbus*. New York, NY: Knopf.

Nevle, R. J., & Bird, D. K. (2008). Effects of syn-pandemic fire reduction and reforestation in the tropical Americas on atmospheric CO_2 during European conquest. *Palaeogeography, Palaeoclimatology, Palaeoecology*, *264*, 25–38.

Oosthuizen, S. (2017). *The Anglo-Saxon Fenland*. Oxford: Wingather Press.

Phillips, J. D. (2001). Human impacts on the environment: Unpredictability and the primacy of place. *Physical Geography*, *22*, 321–332.

Smith, M. W., Macklin, M. G., & Thomas, C. J. (2013). Hydrological and geomorphological controls of malaria transmission. *Earth-Science Reviews*, *116*, 109–127.

Starkel, L. (1987). Man as a cause of sedimentological changes in the Holocene. *Striae*, *26*, 5–12.

Vavilov, N. I. (1951). The origin, variation, immunity and breeding of cultivated plants (K. Starr Chester, Trans.). *Chronica Botanica*, *13*, 1–366.

4

EARTH, AIR, FIRE AND ICE: EXPLORING LINKS BETWEEN HUMAN-INDUCED GLOBAL WARMING, POLAR ICE MELT AND LOCAL SCALE EXTREME WEATHER

Edward Hanna and Richard J. Hall

Global temperature has risen by around 1 °C since 1900 (Fig. 4.1a). This might not sound a great deal (if you change by 1 °C the temperature of the room you're sitting in, you would barely notice) but – because temperature statistics are based on averaging thousands of weather stations' data worldwide – this is a highly significant increase when compared with previous centuries going back for at least 1,000 years and probably much longer. Also, global warming has generally accelerated within the last few decades, so that at least half of the last century's increase is since around 1970. Nearly all climate scientists agree that the main cause of this temperature rise is ever-increasing accumulations of 'greenhouse gases', especially carbon dioxide and methane, within

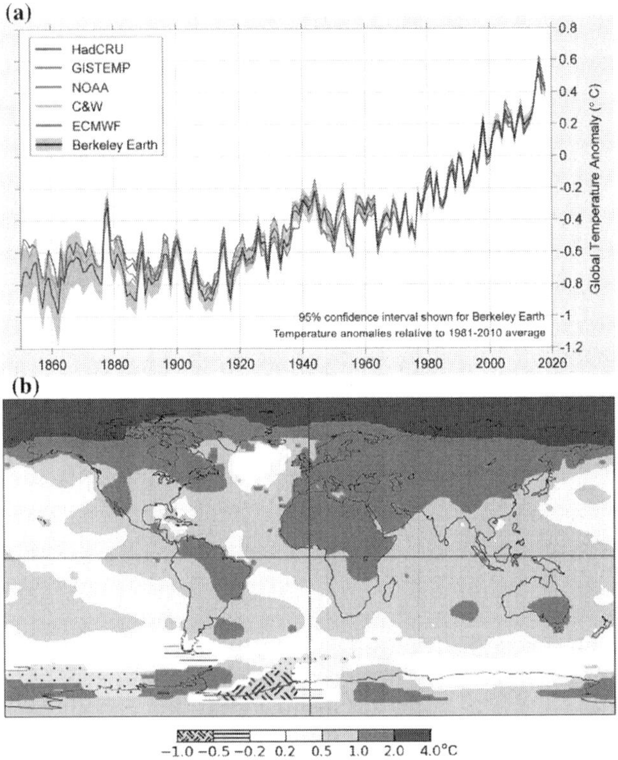

Source: (a) Reproduced from http://berkeleyearth.org/2018-temperatures/. (b) Based on NASA/GISS data available from https://data.giss.nasa.gov/gistemp/maps/index_v4.html (GISTEMP Team, 2020).

Fig. 4.1. (a) Average Global Near-surface Temperature for 1850–2018, Relative to 1981–2010, from Several Independent Reconstructions. The grey shading in the graph indicates uncertainties. (b) Global Temperature Anomalies for the Decade 2009–2018, Relative to 1951–1980 Average Global Temperatures. Dotted regions are areas of no data. Note the Arctic Amplification signal across the top of the map.

our atmosphere. At the time of our Lincoln Moana conference (September 2019), the global atmospheric concentration of carbon dioxide stood at 408.54 parts per million (ppm) by volume, which has increased from 313.2 ppm in September 1958, the year when continuous global monitoring began (https://www.esrl.noaa.gov/gmd/webdata/ccgg/trends/co2_data_mlo.pdf). Some 3.03 ppm of this increase occurred in the last year of record since September 2018, which illustrates the accelerating nature of rising greenhouse gas levels that currently show no sign whatsoever of abating. Just to give a little more context, over most of the last 10,000 years (Holocene interglacial period) before the Industrial Revolution, atmospheric concentrations of greenhouse gases were relatively remarkably stable at around 280 ppm, although they dipped to 180–200 ppm during deep glacial periods such as the Last Glacial Maximum about 18,000–20,000 years ago (Eggleton, 2013). Observational data indicate that current levels of atmospheric carbon dioxide are unprecedented in the last few million years (Zhang, Pagani, Liu, Bohaty, & DeConto, 2013). The second most important greenhouse gas in the atmosphere that is directly influenced by human activity and is contributing to global warming is methane, which has similarly shown a relentless upward trend over the last few decades (https://www.esrl.noaa.gov/gmd/ccgg/trends_ch4/).

Much of the authors' research over the last 20 years has been focused on Arctic climates and the mass balance of the Greenland Ice Sheet – a giant repository of ice and snow whose surface area encompasses the size of several of the larger European nations (approximately the combined size of France, Spain, Sweden and the United Kingdom) that is largely a relict feature from the last ice age and should not really exist according to present-day climate (Ridley, Gregory, Huybrechts, & Lowe, 2009): it is currently maintained through a series of delicate 'feedback' balances. This huge mass of ice is therefore already living on borrowed time and, if it were to melt in its entirety, would contribute some 7 metres to global sea level. Admittedly the best available computer model simulations suggest that, even under conditions of sustained global warming, removing the Greenland Ice Sheet would take more than 1,000 years in its entirety (Ridley et al., 2009). However, the important point here is that this process has already begun; from 2003 to 2018 the ice sheet lost at least 250 billion tonnes of mass each year to the oceans, with a three to four times increase in its mass-loss rate during this period (Hanna et al., 2020). Even more significant are the Antarctic Ice Sheets which store a mass equal to some 65 metres of sea level rise and are currently also losing mass, somewhere in the region of 150 billion tonnes per year. There is recent evidence that this 'sleeping giant of Antarctica' is slowly awakening, with the possibility of many tens of centimetres global sea level rise should the West Antarctic Ice Sheet become destabilised, as some scientists have predicted. Indeed this process may arguably already have begun; even though the Antarctic climate is generally much colder than Greenland, there are crucial regions – especially around the Antarctic Peninsula – which have also been warming relatively rapidly.

Melting ice on land, in the form of wasting glaciers and ice sheets, raises sea level, as does the thermal expansion of warmer sea water, but what about sea ice – the frozen lid formed when parts of the high-latitude oceans freeze each winter? Arctic sea ice has shrunk by about 50% in area and three-quarters in volume in September since continuous satellite monitoring began in the late 1970s, with a record low area year in 2012 (Strove & Notz, 2018), and 2019 being among the lowest years recorded (Fig. 4.2). The present low level of late summer Arctic sea ice cover has not been seen for at least several thousand years. Assuming that recent trends

(a)

(b)

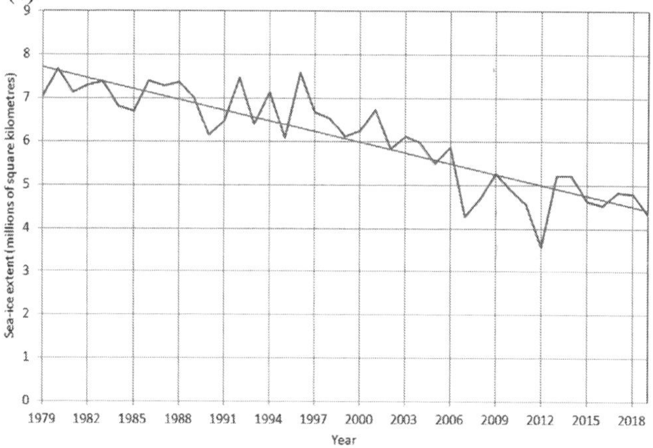

Source: (a) Photographed by E. Hanna on 27 July 2019. (b) Based on data from the US National Snow and Ice Data Center (NSIDC, 2020).

Fig. 4.2. (a) Sea Ice Floes in the Beaufort Sea, North of Alaska. (b) Time Series of Northern Hemisphere Sea Ice Extent Anomalies in September (the Month of Minimum Ice Extent) for 1979 to 2019. The anomaly value for each year is the difference (in percent) in ice extent relative to the mean values for the period 1981–2010. The linear regression line indicates ice losses of −12.8% per decade, which are statistically significant at the 99% confidence level.

continue, which seems more than likely given the increased solar energy absorbed by a barer ocean surface and other so-called 'positive feedbacks' that tend to accelerate this kind of decline, we are currently on course to remove essentially all the sea ice in the Arctic Ocean before – possibly well before – 2050. Sea ice melt does not contribute to sea level rise but it is a major factor in warming the Arctic.

The Arctic has recently experienced an accelerated warming of about double the average rate of global warming (Fig. 4.1b). This polar or Arctic amplification (Overland et al., 2019) is caused by several 'feedback' processes involving the loss of snow and ice cover that accelerate the initial warming and snow/ice melt. One of them is called the albedo (reflection) feedback, in which since snow and ice are typically bright, white and highly reflective, their progressive removal darkens the surface, which results in more energy being absorbed from the Sun, which in turn increases surface warming and exacerbates melt. Some of this darkening involves microbes and algae in the ice, which have relatively low reflectivity and become more active with progressive warming. This is a graphic example of another life-based process helping to accelerate ice melt and regional/global warming (Cook et al., 2020). Another feedback involves newly exposed/ice-free ocean surface in autumn/early winter releasing a much greater amount of heat to the (much colder) air above, once the highly insulating sea ice blanket has been removed or sufficiently thinned or broken (Fig. 4.2).

Global warming is not just about rising temperatures, melting ice and sea level rise, but it also affects the frequency and severity of many extreme weather events. 2019 'Geo Risks' data from the global reinsurance group, Munich Re, show a fourfold increase from 1980 to 2018 in the number of meteorological, hydrological and climatological events that caused at least one fatality and/or produced losses above some pre-defined threshold (adjusted according to the World Bank income group of the affected country), even though geophysical events (earthquakes, tsunamis and volcanos) remained more or less constant in terms of their human and economic impacts during this time period. Just taking a few recent examples, 2019 saw a new all-time UK record temperature of 38.7 °C (Cambridge Botanic Garden, 26 July; UK Met Office), while new temperature records were also set in several European countries including 45.9 °C in France on 28 June (Metéo France). French health industry data indicate that 1,465 extra deaths – a 9% increase on the seasonal average – occurred during the June and July 2019 heatwaves (although, due to preventative measures, this was much lower than equivalent casualties during the infamous 2003 European summer heatwave). Air from the European heatwave eventually found its way over Greenland where it resulted in a new record temperature of 1.1 °C at Summit (right in the centre of the country, some 3,200 metres above sea level) on 31 July 2019 (NSIDC, 2019). Also, Australia also experienced its hottest two summers on record in 2018–2019 and 2019–2020 (Australian Bureau of Meteorology). Severe wildfires ravaged many parts of Europe from Portugal to Greece to the United Kingdom (Saddleworth Moor near Manchester) during a record dry, warm period in summer 2018 and southern Australia during December 2019/January 2020.

Warmer air is also able to hold more water vapour which results in heavier rainfall in some regions, even while other areas, such as the subtropics, tend to become drier with shifting patterns in air circulation. So while, for example, June and July 2019 saw extreme heat in parts of Europe, there was also heavy rain and floods in certain locations such as in central and eastern England in June and this was one of the causes of the Wainfleet, Lincolnshire, floods on 12 June. In England and Wales, the last 5–15 years has seen some record wet summer episodes, most notably in June/July 2007, when large parts of Hull and Sheffield were underwater with several deaths directly attributed to these floods, and again in 2012, as well as some extreme autumn rains and resulting flooding in November 2009 (Lake District) and November 2019 (focused in South Yorkshire, the East Midlands and Lincolnshire). The record wet spring/summer 2012 was especially interesting as it came after an unusually dry period in 2011/ early 2012 in England/Wales, indicating possible increased variability of extremes and water availability.

Winter in the United Kingdom has recently seen extreme weather at opposite ends of the spectrum. 26 February 2019 experienced the United Kingdom's warmest winter day on record, with 21.2 °C recorded at Kew Gardens in London, while almost exactly a year earlier, 1 March 2018, during the famous 'Beast from the East' meteorological episode, was widely the coldest March day across England in over a century (UK Met Office). These severe events were mirrored by similar weather extremes during the last decade, including record wet and stormy spells in winters 2013/14 (when the Thames widely flooded near London) and 2015/16 (focused further north in England, with central York floods), and severe cold and snow in winter 2009/10 and again in December 2010. The latter month was the coldest December in the Central England Temperature record since 1890, while December 2015 was the warmest Central England December ever recorded, with an average temperature of more than 5 °C above the long-term December average and almost no frosts during the entire month (UK Met Office). These greater UK winter extremes since 2010 result from a doubling of jet stream variability across the North Atlantic Ocean (Fig. 4.3). The jet stream is a giant current of air that broadly flows from west to east over mid-latitude regions around the globe. In the North Atlantic, the jet stream steers the low-pressure systems

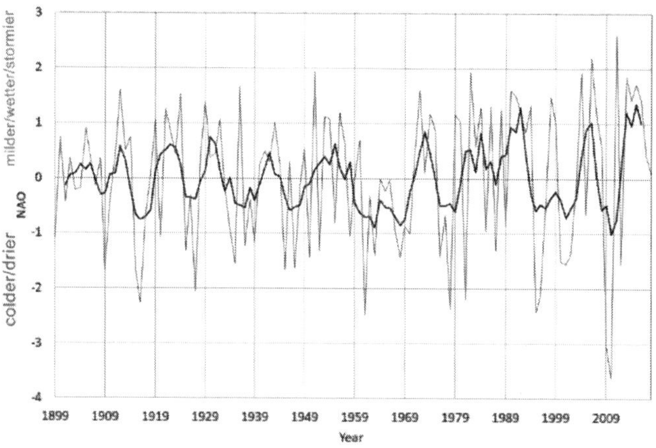

Source: Updated from Hanna, Cropper, Jones, Scaife, and Allan (2015).

Fig. 4.3. Analysis of the North Atlantic Oscillation (NAO) Index, i.e. the Strength of Westerly Winds Reaching the United Kingdom. The graph shows December NAO values from 1899 to 2018. NAO data are derived using Dr James Hurrell's principal component-based index, archived at https://climatedataguide.ucar.edu/climate-data/hurrell-north-atlantic-oscillation-nao-index-pc-based.

which bring rain to the United Kingdom and Western Europe. It tends to point on average towards the United Kingdom, but this increasing variability means that the storms can be steered further north or south more frequently. Various factors, such as North Atlantic sea temperature changes, the El Niño ocean-climate system originating in the tropical Pacific, melting Arctic sea ice and changes in the amount of energy coming from the Sun, seem to have conspired to make the winter jet significantly more variable over the United Kingdom region, although this is one important aspect of climate research that is not properly understood and urgently requires more work.

Another major part of the authors' recent work has focused on testing a scientific hypothesis that this strong Arctic warming since the early 2000s may have destabilised the jet stream further south that controls weather variations and extremes that directly affect hundreds of millions of people who live within the Northern Hemisphere mid-latitudes, including large parts of Eurasia as well as North America (Francis & Vavrus, 2012; Overland et al., 2015). The essential idea is that, since the north–south gradient in temperature is a major source of energy for driving the jet stream, if the polar regions warm faster than elsewhere, the result is a reduction in the efficiency of this 'heat engine'. Of course, the real atmosphere is much more complex than this simple hypothesis suggests. However, there is evidence from meteorological observations that the North Atlantic jet stream has indeed become weaker and wavier in summer in the last 10–15 years compared with the previous few decades. This change has been accompanied by increase in a phenomenon called Greenland high-pressure blocking, which represents areas of high pressure that remain nearly stationary over the Greenland region and distort the usual progression of storms across the North Atlantic. This 'blocking' is linked to some of the most extreme UK seasonal weather events experienced since 2007 (e.g. Hanna, Hall, & Overland, 2017; Overland, Francis, Hanna, & Wang, 2012). It is also thought to be linked to weather extremes elsewhere such as in October 2012 Hurricane Sandy, that due to high blocking over and south of Greenland, took a highly unusual westerly track into New York, rather than recurving out eastwards across the North Atlantic (Mattingly, McLeod, Knox, Marshall Shepherd, & Mote, 2014). The most extreme Greenland blocking summers, 2012 and 2019, saw record surface temperatures and melt across the ice sheet and were also notable for their extreme weather downstream over Northwest Europe, including the United Kingdom. These are just a few examples of where weather events in different regions influence each other through changes in the jet stream that can in turn be affected by human-driven climate change. The key message is that global warming is not a uniform process and can spring surprises in regional climate change and is probably linked with the tendency for Northern Hemisphere mid-latitudes to have more extreme (variously hot/cold/dry/wet) weather, especially during the recent period (since 2007) of rapid Arctic warming. Furthermore, as we have seen, there is also the serious question of sea-level rise arising from ice sheet and glacier melt. We ignore what is happening in remote polar regions at our peril!

So, what about the future? The level of global warming and resulting impacts that we have so far experienced are relatively modest compared with even mid-range predictions of the amount of extra changes that we can expect by the end of the century (Fig. 4.4). Under a high greenhouse gas emissions (business as usual) scenario, the end of century (2080–2099 average) warming is likely to be 3.2–5.4 °C above pre-industrial temperature levels (IPCC, 2019). These are high levels of warming with very serious climate impacts worldwide. For example, although this is near the upper end of the predicted range, there is a distinctly 'likely' possibility of having more than 1 metre of global sea level rise by 2100 if greenhouse gas emissions continue largely unchecked. The Intergovernmental Panel on Climate Change (IPCC, 2018) showed the importance of limiting human-induced global warming to 1.5 °C above pre-industrial. There are significant benefits, for example, to ice sheet stability and sea level rise, but many other aspects also, if the overall amount of global warming is limited to 1.5 °C compared with 2 °C (e.g. Pattyn et al., 2018). It is important to remember here that since 1900 global temperatures have already risen by 1.0 °C due to human influences and are currently on track to breach the 1.5 °C limit as early as 2030 and very likely before 2050.

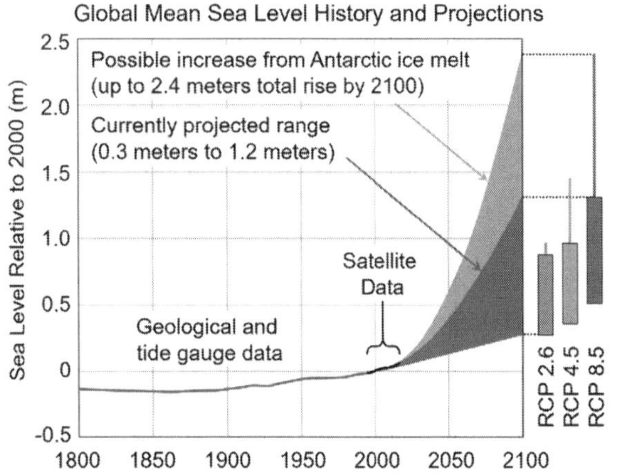

Source: Published by the US Global Change Research Program for the Fourth National Climate Assessment (Sweet et al., 2017).

Fig. 4.4. Historical Sea Level Reconstruction and Projections up to 2,100. The boxes to the right show the likely ranges in sea level rise by 2100, taking a year 2000 baseline, that correspond to the different Research Concentration Pathway (IPCC) scenarios. RCP2.6 is the scenario where emissions peak before 2020, RCP 4.5 the one where they peak around 2040, and RCP8.5 the one where they keep increasing as usual. The lines above the boxes show possible increases based on recent research on the potential Antarctic Ice Sheet contribution to sea level rise (see Hanna et al., 2020, Section 3.2 and their Fig. 5, for further discussion).

Limiting global warming to 1.5 °C above pre-industrial temperatures requires global carbon dioxide emissions to decrease by approximately 40–60% by 2030 relative to 2010 levels, compared with reducing such emissions by 10–30% over the same time frame to limit warming to 2 °C (IPCC, 2018). This is clearly a massive but urgent challenge that requires widespread and deep emissions reductions. A glance at the accelerating increase of atmospheric carbon dioxide over the last 60 years does not show any sign yet of a slow down. Since this report was produced, and based on the recommendations of the UK Committee on Climate Change (CCC, 2019), the United Kingdom on 27 June 2019 became the first major economy that committed to reducing its greenhouse gas emissions to net zero by 2050 (which was updated from a previous target of at least 80% reduction compared with 1990). However, this kind of intention is required internationally and, even more crucially, needs to be fully followed through by setting suitable interim goals and achieving these through appropriate actions. Such follow-up must be fully embraced at all levels of government and industry, as well as amongst the wider public where – although general awareness has recently increased (for example, witness Greta Thunberg and Extinction Rebellion protests in 2019) – there remains big differences in opinion and probably still a great deal of collective 'cognitive dissonance'.

Subsequent to the IPCC (2018) and CCC (2019) reports, there was the declaration of climate emergencies by multiple UK institutes, including the University of Lincoln, United Kingdom, on 16 May 2019, and the establishment of local government working groups, such as the Lincoln Climate Commission: these are essential first steps for tackling these issues at local to regional levels, but, equally clearly, a lot more work will be needed, for example, regarding the more rapid adoption of renewable energies in a world where energy consumption has doubled since 1980 (Leatherby & Martin, 2019). We are currently at a critical cusp in an ongoing race between climate change and market economy forces that have recently promoted the accelerated use of cheaper clean energy sources against a backdrop of increased overall energy demand. No one yet knows whether and how quickly the climate battle will be won but, apart from winning more collective hearts and minds, business and economic solutions (which ultimately influence much of everyday human behaviour) are crucial for solving the problem.

In summary, there is overwhelming scientific evidence that human activity through enhanced greenhouse gas emissions is largely responsible for recent climate change and accompanying extreme weather, and we are already clearly seeing these changes; although some regional aspects of the process remain unclear, the basic science is not in doubt. However, it's equally evident that, although initial remedial steps are being taken, finding an adequate solution will not be easy unless much larger changes are made to the way in which we all live. It's a huge challenge, but we now have the knowledge and technology to meet it, just as we did with the Montreal Protocol in 1988 for the Antarctic ozone hole crisis (albeit for a much simpler problem), and so a successful solution lies in our collective hands. However, there is no mistaking that it will require a huge amount of political and social willpower, and an unprecedented focus and coordination across all scales, to achieve. There will need to be a collective solution that fully involves diverse communities, including religious stakeholders, which on a global scale will mean investing much more in renewable energy and switching from coal and gas to cleaner sources of electricity including nuclear, in order to 'decarbonise' the economy. It's important to remember that this is not someone else's problem; it's a problem that affects all of us and the generations to come. Religious communities are particularly well placed to help lead such a change because of their significant influence on hundreds of millions of people in an age where trust in conventional political systems has, to a large extent, broken down, and also due to their sizeable financial investments which provide a very useful and timely opportunity to go much further down the carbon divestment (withdrawal of support of fossil fuel industries) pathway.

REFERENCES

Committee on Climate Change. (2019, May). Net zero – The UK's contribution to stopping global warming. CCC report. Retrieved from https://www.theccc.org.uk/publication/net-zero-the-uks-contribution-to-stopping-global-warming/

Cook, J. M., Tedstone, A. J., Williamson, C., McCutcheon, J., Hodson, A. J., Dayal, A., … Tranter, M. (2020). Glacier algae accelerate melt rates on the south-western Greenland Ice Sheet. *The Cryosphere*, *14*, 309–330.

Eggleton, T. (2013). *A short introduction to climate change*. New York, NY: Cambridge University Press.

Francis, J. A., & Vavrus, S. J. (2012). Evidence linking Arctic amplification to extreme weather in mid-latitudes. *Geophysical Research Letters*, *39*, 1–6, L06801.

GISTEMP Team. (2019). *GISS Surface Temperature Analysis (GISTEMP)*. NASA Goddard Institute for Space Studies. Retrieved from data.giss.nasa.gov/gistemp/. Accessed on November 8, 2019.

Hanna, E., Cropper, T. E., Jones, P. D., Scaife, A. A., & Allan, R. (2015). Recent seasonal asymmetric changes in the NAO (a marked summer decline and increased winter variability) and associated changes in the AO and Greenland Blocking Index. *International Journal of Climatology*, *35*, 2540–2554.

Hanna, E., Hall, R. J., & Overland, J. E. (2017). Can Arctic warming influence UK extreme weather? *Weather*, *72*, 346–352.

Hanna, E., Pattyn, F., Navarro, F., Favier, V., Goelzer, H., van den Broeke, M., … Smith, B. (2020). Mass balance of the ice sheets and glaciers – Progress since AR5 and challenges. *Earth-Science Reviews*, *201*, 102976.

IPCC. (2018). Summary for policymakers. In V. Masson-Delmotte, P. Zhai, H.-O. Pörtner, D. Roberts, J. Skea, P. R. Shukla, … T. Waterfield (Eds.), *Global warming of 1.5°C. An IPCC special report on the impacts of global warming of 1.5°C above pre-industrial levels and related global greenhouse gas emission pathways, in the context of strengthening the global response to the threat of climate change, sustainable development, and efforts to eradicate poverty* (32 pp.). Geneva: World Meteorological Organization.

IPCC. (2019). Summary for policymakers. In H.-O. Pörtner, D. C. Roberts, V. Masson-Delmotte, P. Zhai, M. Tignor, E. Poloczanska, … N. Weyer (Eds.), *IPCC special report on the ocean and cryosphere in a changing climate*. Intergovernmental Panel on Climate Change. Retrieved from https://report.ipcc.ch/srocc/pdf/SROCC_SPM_Approved.pdf

Leatherby, L., & Martin, C. (2019). How each country contributed to the explosion in energy consumption. Retrieved from https://www.bloomberg.com/graphics/2019-international-energy-use-renewables-coal-oil/

Mattingly, K. S., McLeod, J. T., Knox, J. A., Marshall Shepherd, J., & Mote, T. L. (2014). A climatological assessment of Greenland blocking conditions associated with the track of Hurricane Sandy and historical North Atlantic hurricanes. *International Journal of Climatology*, *35*, 746–760.

NSIDC. (2019). Europe's warm air spikes Greenland melting to record levels. Greenland Ice Sheet Today. National Snow and Ice Data Center. Retrieved from http://nsidc.org/greenland-today/2019/08/europes-warm-air-spikes-greenland-melting-to-record-levels/

NSIDC. (2020). Sea ice index Arctic- and Antarctic-wide changes in sea ice. Retrieved from https://nsidc.org/data/seaice_index/archives. Accessed on March 1, 2020.

Overland, J. E., Francis, J. A., Hall, R. J., Hanna, E., Kim, S. J., & Vihma, T. (2015). The melting Arctic and midlatitude weather patterns: Are they connected? *Journal of Climate*, *28*, 7917–7932.

Overland, J. E., Francis, J. A., Hanna, E., & Wang, M. (2012). The recent shift in early summer Arctic atmospheric circulation. *Geophysical Research Letters*, *39*, 1–6, L19804.

Overland, J. E., Hanna, E., Hanssen-Bauer, I., Kim, S.-J., Walsh, J. E., Wang, M., … Ballinger, T. J. (2019). Surface air temperature in Arctic report card 2019. Retrieved from https://arctic.noaa.gov/Report-Card/Report-Card-2019/ArtMID/7916/ArticleID/835/Surface-Air-Temperature

Pattyn, F., Ritz, C., Hanna, E., Asay-Davis, X., DeConto, R., Durand, G., … van den Broeke, M. (2018). The Greenland and Antarctic ice sheets under 1.5°C global warming. *Nature Climate Change*, *8*, 1053–1061.

Ridley, J., Gregory, J. M., Huybrechts, P., & Lowe, J. (2009). Thresholds for irreversible decline of the Greenland ice sheet. *Climate Dynamics*, *35*, 1049–1057.

Strove, J., & Notz, D. (2018). Changing state of Arctic sea ice across all seasons. *Environmental Research Letters*, *13*, 1–23, 103001.

Sweet, W. V., Kopp, R. E., Weaver, C. P., Obeysekera, J., Horton, R. M., Thieler, E. R., & Zervas, C. (2017). *Global and regional sea level rise scenarios for the United States*. Technical Report NOS CO-OPS 083. National Oceanic and Atmospheric Administration, Silver Spring, MD.

Zhang, Y. G., Pagani, M., Liu, Z., Bohaty, S. M., & DeConto, R. (2013). A 40-million-year history of atmospheric CO_2. *Philosophical Transactions of the Royal Society A*, *371*(2001), 20130096.

5

CLIMATE CHANGE AND THE ROLE OF EDUCATION

Sarah Hemstock, Siu Fanga Jione, Mark Charlesworth, and Patrina Dumaru

INTRODUCTION

We are living in the Anthropocene, and climate change is the major contemporary societal challenge. In order to help people comprehend the various messages around climate change, and indeed, in order to help society at large adapt and survive, it is important to foster climate change education. Education's key role in addressing climate change and other environmental issues was initially recognised in international policy at the United Nations Conference on Environment and Development (UNCED), also known as the Rio de Janeiro Earth Summit in 1992. The United Nations Framework Convention on Climate Change (UNFCCC) was also opened at the Earth Summit.

Pacific Small Island Developing States (PSIDS) are on the 'frontline' of climate and environmental change impacts and are also at the forefront of developments in climate change education. Current Pacific regional and national policy strongly supports climate change education (Leal Filho & Hemstock, 2019). This chapter explores the international policy context which has led to groundbreaking Pacific educational initiatives before considering a PSIDS case study.

AGENDA 21

Agenda 21 (UNCED, 1992) set the overall UN framework for environment and development questions in terms of sustainable development at the Rio Summit. The document had a chapter dedicated to education and mentioned the term 250 times. This includes mention of spiritual development (para. 36.3), and a recognition of urgency, with timescales of 'three years' (para. 36.5). Of particular relevance is the call to reinforce 'values that support sustainable consumption ... organizations should promote more positive attitudes towards sustainable consumption through education' (para. 4.26).

UNFCCC

The UNFCCC (1992) placed less emphasis on education as Agenda 21 was seen as the context in which the UNFCCC would operate. However, UNFCCC Article 4 addresses education directly: 'Promote and cooperate in education, training and public awareness related to climate change and encourage the widest participation in this process'. Article 6 covers 'Education, Training and Public Awareness' more broadly with 'public awareness programmes', 'access to information', 'public participation in addressing climate change' and, with relevance to the recent Pacific vocational sector developments, 'training of scientific, technical and managerial personnel'. The success of 'public participation in addressing climate change' appears limited with policymakers appearing to prefer to emphasise experts delivering economic analysis and technological developments rather than engaging citizen questions of sustainable consumption or involving them directly in policy processes. The lack of engagement in education more widely by the UNFCCC is also confirmed by the establishment of the Paris Committee on Capacity Building. Their first meeting in May 2017 was 25 years in the making, and disappointingly directed UNFCCC educational efforts towards 'training' countries to produce better evaluations of 'nationally determined contributions', rather than recognising climate change as an employment sector and providing a means of professionalising that sector, as was lobbied for by PSIDS such as Vanuatu and the Solomon Islands (PCCB, 2017).

SUSTAINABLE DEVELOPMENT GOALS

The 2015 Sustainable Development Goals discussion of climate change did not mention education specifically, and discussion of education did not include climate change, but this is a brief document where it is reasonable to suggest the Agenda 21 and UNFCCC framings are implicit. Additionally, SDG4 is focused on 'quality education'.

SENDAI FRAMEWORK

In contrast to the international frameworks and policies outlined above, the Sendai Framework for Disaster Risk

Reduction 2015–2030 (Sendai Framework) has a clear and practical guide to achieving its goals, and capacity development via education is key. Its approach to education is laid out in the United Nations Office for Disaster Risk Reduction (UNDRR) 'Strategic Approach to Capacity Development for Implementation of the Sendai Framework for Disaster Risk Reduction' (UNDRR, 2019). Since the PSIDS are the most vulnerable countries to natural hazards, and they are leading in the development of vocational sector DRR qualifications, they were widely consulted during the development of this strategy. Increasingly practitioners, development funders and policy makers are following the Pacific countries lead by linking disaster risk management (DRM) and climate change adaptation (CCA) (Hemstock et al., 2018). The Pacific understanding of climate change as a slow-acting disaster was adopted by the European Union Pacific Technical Vocational Education and Training on Sustainable Energy and Climate Change Adaptation Project (EU PacTVET – a €6.1 million project currently being implemented in 15 PSIDS by the Pacific Community) in introducing innovative initiatives to address wide-ranging capacity needs (EU PacTVET, 2016, 2017).

PACIFIC CASE STUDY

For many PSIDS, the key barrier to improving national resilience to disaster risk and climate change impacts was identified as a lack of capacity, resulting from the absence of sustainable accredited and quality assured formal training programmes in DRM and CCA (Buliruarua et al., 2015). Buliruarua et al. (2015) revealed that formal qualifications were required for 'genuine' capacity building with 11 PSIDS identifying some component of quality assessment (qualification and delivery accreditation and institutional validation) as key areas for assistance. For example, a country such as Tuvalu with a population of just 11,000 people would struggle to develop and accredit national qualifications, and it would be unlikely that a Tuvalu national qualification would be recognised in other countries. Therefore, a regional (as opposed to national) accreditation and quality assurance mechanism for formal qualifications and institutional validation makes sense and was developed under the EU PacTVET project. This strategy, a global first for vocational education, allows national delivery of regionally developed and quality assured qualifications in 'Resilience (CCA and DRM)' (EU PacTVET, 2016) and 'Sustainable Energy' (Pacific Community, 2017).

High-level national stakeholders from 15 PSIDS were keen to ensure opportunities for formal qualifications were open to all people affected by climate change, so qualifications were designed in blocks so that they could be made available for everyone, from grassroots community members, through to technicians and government and private sector managers. The qualifications were constructed around a 'competency' and 'skill set' approach. This means that people can pick what competencies they need to 'up-skill' in order to improve their own capacity, achieve a community project, improve food security, ensure clean drinking water, etc. (EU PacTVET, 2016). For example, at levels 1 and 2 'strands' include 'Climate Change and Disaster Risk Reduction' and tackle generic skills such as communication and cultural protocols; workplace health and safety; team work; and work-based practices. Levels 3 and 4 look at the application of knowledge, planning and management in the specialist areas of agriculture; forestry; coastal management; tourism; water resources; energy and infrastructure; fisheries and health. Countries can deliver different aspects of the qualifications (i.e. competencies, which build into skill sets, which build into a qualification) according to their own workforce needs. Additionally, as qualifications are regional, skill sets are mutually recognised and can be built upon by completing competencies/skill sets at more than one educational provider. This is another advantage of this approach since some providers are schools with few facilities and where staff are only able to teach limited strands of the qualifications.

This formal educational approach is a response to the millions of dollars spent in the region on ad hoc, informal, one-off and uncredited trainings in sustainable energy and CCA. Offering formal qualifications supports the advancement of the trainees in their work places and careers. It also supports labour mobility within and outside of the region – this is important since climate change–induced migration is already a reality for some PSIDS communities. Donor support for non-formal training delivered by external providers is not the best use of scarce resources, it is unsustainable as it only lasts for the lifetime of the project and should be discouraged. Genuine capacity building via formal accredited qualifications in 'Resilience' delivered by in-country providers is both sustainable and owned by the communities they serve. Donors need to be encouraged to support community scholarships for the various formal qualification skill sets for any project that has a training element, rather than continuing with their current approach. The sustainability of the formal education approach is assured by the recognition of 'Resilience' as an employment sector for PSIDS and the establishment of an industry association – The Pacific Regional Federation of Resilience Professionals, which was launched at the UNDRR Pacific DRR Platform in 2016 (EU PacTVET, 2017; Hemstock et al., 2018).

SO HOW DOES THIS WORK ON THE GROUND?

Various toolkits have been developed to assess community vulnerability and possibilities for adaptation (Hemstock, McNamara, & Holland, 2013). Several of these are available as learning resources for the 'Resilience' qualifications.

Integrated (multisector) Vulnerability Assessments (IVAs) are important as they help PSIDS develop their National Adaptation Plans which are required to access UNFCCC funding. The IVA Framework (developed in 2016) systematically examines how environmental and developmental changes affect local communities with regard to meeting basic needs (Dumaru et al., 2017). It provides standardised baseline data about community vulnerability that can be periodically replicated to inform the implementation and effectiveness of the National Adaptation Plan (Dumaru, 2019).

In 2015, the 'Community Mapping & QGIS Disaster Risk Mapping Toolkit' was developed to provide communities with a tool for easy data collection, analysis and representation of areas at risk in order to support decision-making through participatory risk mapping (Pouvalu, 2015). Its design promotes community-led initiatives to produce, own and use geographical data for decision-making based on their experiences, priorities and values (Pouvalu, 2015; Sieber, 2006). The toolkit consists of the QGIS software (v.2.6.0 at the time of publication), a step-by-step manual, spatial data and video tutorials to guide users on data collection, management, analysis and representation using the QGIS software.

The GIS toolkit was designed to operate on its own; however, its participatory nature gave it flexibility to be used to complement and/or justify the findings of other participatory approaches such as the IVA. Between 2016 and 2018, six regional training workshops in Tuvalu, Nauru, Tonga,

Fiji and Vanuatu were conducted to train participants on the use of the IVA and GIS toolkit. Participants consisted of community members, practitioners and representatives from the government and NGOs. The workshops were conducted in two phases. The first focused on contextualising the IVA to issues faced by each country prior to assessing the vulnerabilities of the local communities. The second phase focused on using the GIS toolkit to visualise the areas at risk based on the results of the IVA. A common observation throughout the workshops was the change in perception of the participants pre and post assessment. What participants initially perceived to be the areas to prioritise differed from the actual findings. For instance, in Tuvalu, participants thought that water security needed to be prioritised by providing more water storage tanks. However, the IVA results indicated that water security was influenced by governance at the household level. The assessments suggested that water supply was sufficient, but the challenge was living in extended families and not being able to share the water resources wisely. Complementing the IVA results, the GIS resource mapping showed that 80% of the households in the community owned one or two water tanks, but more than 50% of the households housed 10–15 people sharing a 2,500-litre water tank. These results indicated that governance at the household level needed strengthening, but the problems of overcrowding also affected household water security and sanitation. Such findings provide communities with a realistic geographical picture of existing vulnerabilities and areas that need strengthening.

In 2017, a 2-week workshop was held in Tonga in response to a request from the Anglican Diocese of Polynesia to provide training to support a youth-led initiative to build resilience within their communities. Participants included youth members from Tonga, Fiji, Samoa and New Zealand

and men, women and members of the clergy from Fasi village in Tonga. Participants were trained on carrying out assessments using the IVA and the GIS toolkit (Diocese of Polynesia, 2017). Men, women and members of the clergy provided historical information of the village, traditional practices, natural resources and experiences. The youth focused on conducting the assessments, data collection and technical aspects of the training. As a result of the training, a member of the clergy stated that he was a priest in Fasi for 10 years and was unaware that a few of the houses surrounding the church had no water supply. He added that the workshop provided specific skills and information around the needs and vulnerabilities of the Fasi community that the church will use to lead and support the community (Jione, 2017).

In 2018, the Anglican youth from Tonga used the skills gained from the training and carried out vulnerability assessments before Tropical Cyclone Gita hit Tonga. Households were categorised according to their level of risk and their locations mapped accordingly. Based on their assessments, 22 households were categorised as 'high risk', and the youth assisted in household reinforcements, strengthening of windows and roofs, provision of supplies and first-aid kits and removal of debris. In the aftermath, the group conducted post-disaster assessments and developed a damage assessment report which was sent to their main office in New Zealand. As a result, two containers of supplies were sent to Tonga based on the information provided in the vulnerability and damage assessments. Supplies were distributed to the affected households based on the assessments and severity of damage (Anglican Tonga, 2018; Jione, 2018). Their experiences were presented by the youth at the Moana – Water of Life: Navigating Climate Change for Planetary Health conference in Lincoln, UK, 2019.

REFERENCES

Anglican Tonga. (2018). All Saints' launches cyclone plan. Retrieved from http://anglicantaonga.org.nz/news/tikanga_pasifika/fasi

Buliruarua, L.-A., Hemstock, S. H., Jacot Des Combes, H., Kua, N., Martin, T., Satiki, V., & Manuella-Morris, T. (2015). *P-ACP training needs and gap analysis. Reports for Fiji, Cook Islands, the Federated States of Micronesia, Kiribati, Marshall Islands, Nauru, Niue, Palau, Papua New Guinea, Samoa, Solomon Islands, Timor Leste, Tonga, Tuvalu and Vanuatu*. Professional and technical report EU PacTVET – FED/2014/347-438. The Pacific Community and The University of The South Pacific, Suva, Fiji.

Diocese of Polynesia. (2017). Vulnerability assessment of Fasi community, Nuku'alofa. Using the Community Integrated Vulnerability Assessment (CIVA) framework and the community mapping toolkit. Retrieved from https://www.dropbox.com/s/jy2arosr8szzed3/CIVA%20%20QGIS%20Training%20Report%20for%20Fasi%20moe%20Afi%20community%20Tonga.pdf?dl=0

Dumaru, P. (2019). How integrated vulnerability assessments support National Adaptation Planning processes in the Pacific region. Newsletter. International Institute for Sustainable Development (IISD), Winnipeg, Canada. Retrieved from http://napglobalnetwork.org/wp-content/uploads/2019/01/napgn-en-2019-how-ivas-support-nap-processes-in-the-pacific-region.pdf

Dumaru, P., Martin, T., Lowry, B., Manuella, T., Koppert, T., Deiye, T., … Southcombe, D. (2017). *Framework guide for community resilient development: Pacific Islands Community Integrated Vulnerability Assessment (CIVA): A tool for resilience management*. Unpublished report. University of the South Pacific, Suva, Fiji. Retrieved from https://www.dropbox.com/s/e0cvyg0qsgo0i2e/Frame%20Work%20-%20Part%201%20of%20the%20CIVA%20Toolkit%20.pdf?dl=0

EU PacTVET. (2016). Regional certificate (1-4) in resilience (Climate Change Adaptation and Disaster Risk Reduction/CCA & DRR). EU PacTVET Project. FED/2014/347-438. Pacific Community (SPC) & The University of the South Pacific (USP), Suva, Fiji.

EU PacTVET. (2017). The Pacific Regional Federation of Resilience Professionals (PRFRP). A professional organization to build capacity in the climate change and disaster risk management sectors in 15 small island developing states in the Pacific – Submission to the Paris Committee on Capacity-Building. SPC, Suva.

Hemstock, S. L., Jacot Des Combes, H., Buliruarua, L.-A., Maitava, K., Senikula, R., Smith, R., & Martin, T. (2018). Professionalizing the "resilience" sector in the Pacific Islands Region: Formal education for capacity building. In S. Klepp, & L. Chavez-Rodriguez (Eds.), *A critical approach to climate change adaptation: Discourses, policies and practices*. Routledge Advances in Climate Change Research. Abingdon: Routledge. ISBN:978-1-13-805629-9

Hemstock, S. L., McNamara, K. E, & Holland, A. E. (2013). *PACE-SD guidebook: Participatory vulnerability and adaptation assessment*. Suva: The University of the South Pacific.

Jione, S. F. (2017). *Community Integrated Vulnerability Assessment (CIVA) and community mapping training evaluation report*. Unpublished report. Pacific Centre for

Environment and Sustainable Development and University of the South Pacific, Suva.

Jione, S. F. (2018). *An analysis of the response to the community mapping and QGIS, a climate and disaster risk mapping toolkit*. Unpublished Master's thesis, Pacific Centre for Environment and Sustainable Development and University of the South Pacific, Suva.

Leal Filho, W., & Hemstock, S. L. (2019). Climate change education: An overview of international trends and the need for action. In W. Leal Filho, & S. Hemstock (Eds.), *Climate change and the role of education*. Climate Change Management. Cham: Springer.

Pacific Community. (2017). Theme: Affordable, reliable and sustainable energy and transport services for all. AGENDA ITEM E5 – Update on training, research and capacity building activities on sustainable energy in the Pacific Island countries and territories. Joint paper by the Pacific Community (SPC) and the University of the South Pacific (USP). Third Pacific Regional Energy and Transport Ministers Meeting, Nuku'alofa, Tonga, 24–28 April 2017. Retrieved from http://prdrse4all.spc.int/sites/default/files/17_energy_resolution_-_final_-_e.pdf

Paris Committee on Capacity Building (PCCB). (2017). Submissions to the Paris Committee on Capacity-Building (PCCB). Solomon Islands and Vanuatu. Retrieved from https://unfccc.int/topics/capacity-building/resources/submissions-to-the-paris-committee-on-capacity-building-pccb#eq-2

Pouvalu, S. (2015). *Community mapping & QGIS. A climate and disaster risk mapping toolkit*. Suva: USP Press.

Sieber, R. (2006). Public participation geographic information systems: A literature review and framework. *Annals of the Association of American Geographers*, *96*, 491–507.

Sustainable Development Goals. (2015). Sustainable development goals. Retrieved from https://www.undp.org/content/dam/undp/library/corporate/brochure/SDGs_Booklet_Web_En.pdf

UNCED. (1992). United Nations Conference on Environment & Development Rio de Janerio, Brazil, 3 to 14 June 1992. New York: United Nations: Agenda 21, Rio Declaration, Forest Principles. Retrieved from https://sustainabledevelopment.un.org/content/documents/Agenda21.pdf

UNDRR. (2019, May). Strategic approach to capacity development for implementation of the Sendai framework for disaster risk reduction – A vision of risk-informed sustainable development by 2030. Concise Guide. United Nations Office for Disaster Risk Reduction (UNDRR). Retrieved from https://www.unisdr.org/files/58211_fullconciseguide.pdf

UNFCCC. (1992). United Nations framework convention on climate change. Retrieved from https://unfccc.int/resource/docs/convkp/conveng.pdf

6

TOWARDS CITIZEN GOVERNANCE FOR CLIMATE CHANGE EDUCATION AND JUSTICE: A SCIENCE–POLICY PERSPECTIVE

Theresa G. Mercer and Andrew P. Kythreotis

The Moana Conference had ambitious aims to identify collective actions that can be taken to navigate weather extremes related to human-induced climate change. Throughout the two-day conference common themes emerged. These included the important role that education must play across society, the importance of local, bottom-up approaches to tackling climate change impacts and the need to protect and help the most vulnerable people in society including the youth that will most keenly feel the impacts of – and legislate for – dangerous climate change in their lifetime. So how can these themes directly translate into more effective climate action? This chapter will discuss how society can be more involved in climate research and policy as a more socially equitable and just way of tackling future climate impacts. The role of education – and its associated institutions – in facilitating this approach is integral to this. Therefore, the first section will discuss previous and contemporary social and political conditions in relation to increased and more equitable and just citizen engagement in climate action in the science–policy domain. The second section then explores how collaborative education approaches through Education for Sustainable Development (ESD) can be ramped up to catalyse increased citizen engagement in climate action. We then conclude with some thoughts on future directions for research in ESD and climate change as a more inclusive and just form of climate governance.

SECTION 1: THE SOCIAL AND POLITICAL CONDITIONS FOR MORE EQUITABLE AND JUST CITIZEN CLIMATE GOVERNANCE AND ACTION

It is clearly evidenced that humans are causing increased global warming leading to significant climate impacts (Allen et al., 2018). Various climate impacts are experienced around the world whether it is the recent flooding experienced in Wainfleet, Lincolnshire or more devastating impacts experienced in Fiji, like Cyclone Winston in 2016. These impacts – to human, physical and biological systems – are getting more extreme and dangerous, possibly irreversible, regardless of geographical location, as the Summary for Policymakers on the 2018 Intergovernmental Panel on Climate Change (IPCC) Special Report on 1.5°C states, 'Some impacts may be long-lasting or irreversible, such as the loss of some ecosystems (high confidence)' (Allen et al., 2018, p. 5). Key public figures agree too. For example, David Attenborough recently highlighted at the 2018 24th Conference of the Parties in Poland,

> *Right now, we are facing a man-made disaster of global scale. Our greatest threat in thousands of years. Climate change. If we don't take action, the collapse of our civilisations and the extinction of much of the natural world is on the horizon*
>
> (Kennedy, 2019).

But what can ordinary citizens do to tackle climate change? There is so much information in the public domain on climate science and many people disagree about the degree in which anthropogenic climate change is affecting our human, physical and biological systems. There are multiple policy responses too, all underpinned by complex technical information which doesn't resonate with the laypeople. This confuses the general public as to what is the best form of climate action. Do we as individuals need to do something in relation to our own environmental behavior? Or is it the role of government? Or both? Do we need to assist the scientists and the government in some way? Where do we start as individual citizens who want to see greater forms of social justice in society, yet feel powerless to catalyse any meaningful scientific or policy change? All this confusion has created more barriers to citizens having the capacity to act on climate change. The authors have discussed many of these barriers to increased citizen involvement in climate science and policy (see Kythreotis et al., 2019), yet complete solutions also elude the so-called experts too.

Policy solutions to climate change have been typically based on the science–policy 'linear model of expertise' where 'truth speaks to power' (Jasanoff & Wynne, 1998). That is, we have two realms where science (truth) is separated from power (government and policy). The scientists (e.g. IPCC) report (speak) the truth – their scientific observations of the climate – to governments, who then decide upon and implement the policy based on that objective, scientific evidence. However, simple and legitimate this linear process may appear, it isn't. Scientists politicise the climate and politicians scientise the politics leading to different interpretations of how best to tackle climate change (see Fig. 6.1).

Hence, more confusion is created within society because we have different forms of politics at play, held by different citizens and groups within society. This climate science–policy complex results in more questions. How effective is the science, and therefore the policy? Is it socially just? As Fig. 6.1 illustrates, citizens (the general public) are at the bottom of this process suggesting that they (we) have less of a role in the science–policy hierarchy. This creates some fundamental questions that bring education to the fore in catalysing new

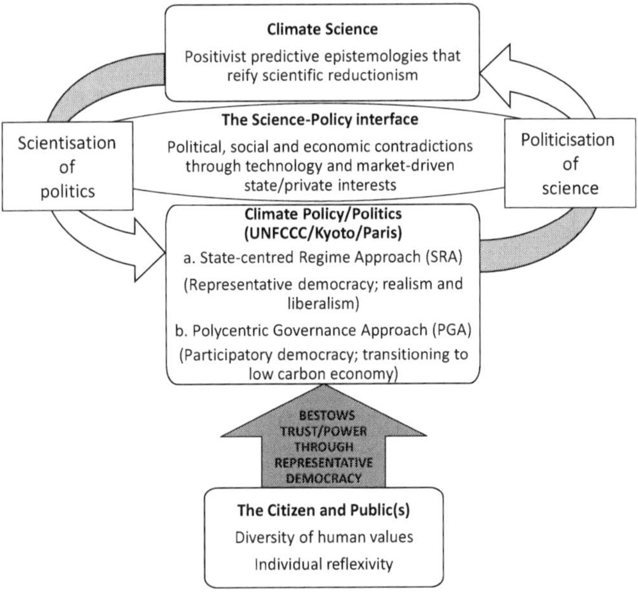

Fig. 6.1 Current Modus Operandi in Formulation of Climate Research and Policy

forms of citizen climate governance and action. Firstly, how do ordinary citizens have a say in climate science and policy? Secondly, should ordinary citizens have a say in climate science and policy? Yes, citizens do have some agency through voting, but if solving climate change relies solely on this, then examples like Donald Trump becoming President of the United States certainly illustrate that democracy may not be conducive to tackling climate change in an equitable and socially just way, and more importantly, at the policy speed required.

People can also collectively demonstrate in public too if they are dissatisfied with their elected governments' inaction on implementing effective climate policy. But can this instil political change and greater engagement of the general public in climate change policy formulation? The recent Extinction Rebellion (XR) movement illustrates that mass civil disobedience demonstrations can affect some policy change, e.g. the call for governments to declare a climate emergency, which has happened. This is all well and good, but are these types of citizen action effective enough, and more importantly quick enough to tackle the increasingly extreme climate impacts we are witnessing through human-induced global warming? We argue no. For example, is XR a legitimate proxy for (rational) wider public opinion on climate change? Yes, XR has given citizen governance some gravitas, e.g. their call for governments to declare a Climate Emergency and introduce Citizen Assemblies to help tackle climate change. But there is another side to XR which could alienate other citizens, e.g. those with more right-winged views (Spicer, 2019).

The recent School Strikes initiated by a 16-year-old Swedish school girl, Greta Thunberg has arguably created certain social and political conditions that have forced us as adults to uncomfortably and soberly re-evaluate what climate change means to future generations. In this sense, maybe *educational institutions* are the climate panacea that we are waiting for, especially given the fact that over 30 years of collective climate negotiations by national governments through the mitigation policy lens have failed to effectively decrease anthropogenic greenhouse gas emissions (Prins & Rayner, 2007).

So, if education is central to tackling climate change, how do we get more citizens involved? Policy has emphasised the importance of education in helping engage citizens with climate change (e.g. World Education Blog, The Commonwealth Education Hub). One example is citizen science where citizens are utilised by scientists to aggregate big data sets to understand environmental change. But citizens are just acting as data collectors really and have no real power in determining new forms of effective climate policy change. Building on the work of Haklay (Haklay, 2013, e.g. 2015), we have argued elsewhere (see Kythreotis et al., 2019) how citizen social science (CSS) – involving greater amount of citizens in the climate science and policy processes – is an important step change needed within this burgeoning era of what we call the 'New Civil Politics of Climate Change.' The next steps for CSS for more effective climate governance and action is though educational institutions – through our schools, colleges, universities *and* churches (regardless of denomination or faith) because these institutions are socially 'good' institutions that are less political in the 'state governing' sense. Thus, they offer more equitable and just recourse for catalysing more effective climate policy because arguably their interests are not underpinned by economic development based on a fossil fuel infrastructure. Section two empirically explores the role of some collaborative education approaches to achieve more effective climate engagement, governance and action.

SECTION 2: THE ROLE OF COLLABORATIVE EDUCATION APPROACHES THROUGH EDUCATION FOR SUSTAINABLE DEVELOPMENT

The idea of embedding climate and environmental issues within education and pedagogic practices is not new. The United Nations Decade for Education for Sustainable Development was introduced in 2005 with the "aim of integrating the principles, values and practices of sustainable development across all aspects of education and learning". In 2015 the Global Action Programme for ESD was launched which builds on the achievements and momentum of the previous Decade's programme. During this time the 17 Sustainable Development Goals were ratified and education (United Nations, 2019) became an important vehicle in achieving all the goals.

Research has shown that younger generations are more attentive to contemporary environmental issues like climate change (Corner et al., 2015) but often lack the necessary institutional avenues to instigate formal political change (Narksompong & Limjirakan, 2015). However, before being able to instigate change, younger people need to be better equipped with the knowledge and skills to be able to make informed decisions and persuade those in positions of policy authority that change is needed. Increased education has been shown to correlate with greater climate awareness (Hess & Maki, 2019). Recent research has shown that when well-informed, children can go on to influence other older family members towards higher levels of climate concern and collective action through intergenerational learning (Lawson et al., 2019). As such there is likely to be a knock-on effect of pro-environmental behaviour from schools-based initiatives throughout wider society.

Both UN programmes recognise the importance for effective ESD across all sectors of education to deliver quality education and to enable sustainable development

(UNESCO, 2014). However, particularly within secondary school teaching, there are issues around the capacity and knowledge of teachers to deliver robust education on sustainability and climate change issues (Bardsley & Bardsley, 2007; McCaffrey & Buhr, 2008). Teachers feel unprepared to teach a subject that is complex, controversial and interdisciplinary in nature (Gayford, 2002; Johnson et al., 2008). This can lead to misconceptions of media and climate scepticism when pupils leave school (Mercer et al., 2017). As such, recent work has highlighted that higher education (HE), and its students can have a potential role in enhancing ESD capacity in primary schools, co-producing climate knowledge and action whilst enhancing their own learning (Mercer et al., 2017). After all, who is better placed? HE students have the appropriate working knowledge, creativity and accessibility to school pupils by being closer to their age.

In the HE sector, 80% of students believe that ESD should be promoted within UK universities with 66% believing it should be incorporated into the curriculum (Drayson, Bone, Agombar, & Kemp, 2014). In recent years there has been an increasing emphasis and mainstreaming of ESD through sustainability awards, green tables, HE environmental strategies, and attempts to embed ESD into degree courses outside of the core environmental disciplines. It has been shown that students studying more environmental-based courses feel the need to educate others in many of the sustainability issues covered on their courses. As such there is scope to include HE students in delivering this ambitious goal of embedding sustainability across universities as well being involved in educating others (Robinson, 2015).

In order to deliver ESD across the educational sector, suitable techniques are needed that shift away from teacher-centred methods to interactive, learner-centred approaches (Cotton & Winter, 2010). The potential for games to be used as a tool for ESD in formal education has been underexploited. Most examples have been used in primary school settings (Pivec, 2007). However, where they have been used, they have been shown to be interactive (Kolb, 1984), contribute to transformational learning (Dieleman & Huisingh, 2006), foster team work and promote interdisciplinarity and knowledge of oneself. Recent work has demonstrated that going beyond the playing of games to game development can be a cumulatively useful learning experience (Mercer et al., 2017). Learning-by-design can make students more motivated and connected with course material.

Recent work on co-production of knowledge sees HE students creating, developing and delivering games to primary school children (Mercer et al., 2017). In this study, environmentally literate students studying for a degree developed educational games related to sustainable lifestyle choices and climate change impacts. Feedback from the primary schools involved with playing the games was largely positive with many pupils taking ideas for sustainability and climate action back to their schools (for example, creating Ecoclubs, holding awareness assemblies etc.) and most school children found the student-created games accessible. The HE students also responded positively to the exercise with most finding improvements to core skills and a request for more of this type of assessment. The exercise also saw enhanced links between the HE institutions and the wider community (Mercer et al., 2017). The idea of co-production of knowledge and action has proved integral to increasing social capacity to initiate change within environmental policy and governance systems (Howarth & Monasterolo, 2016). This type of collaborative approach can enhance greater involvement of younger generations within the climate communication-action-policy process to inculcate wider citizen engagement in climate action through younger generations – which can then potentially catalyse the necessary policy changes at government level.

CONCLUSION

Using examples from our own pedagogic experience, we have argued how collaborative ESD and climate change can be ramped up to catalyse increased citizen engagement in climate governance and action as a form of equitable and just civil politics of climate change. Future research and action in particular should focus on how climate change and sustainability can be embedded across the curriculum in HE settings so that environmental literate students are not just constrained to those studying the obvious environmental/geography focussed courses (Leal Filho et al., 2019; Molthan-Hill, Worsfold, Nagy, Leal Filho, & Mifsud, 2019). Opposition to sustainable policies and interventions can be reduced by improving awareness and belief in climate change through inclusion of these topics in the core curriculum to address this (Hess & Maki, 2019). There also needs to be more provision to allow HE students to co-produce climate knowledge and action across both primary and secondary school settings making the process of climate action more equitable and inclusive. Whilst the study by Mercer et al. (2017) has discussed the application of ESD with primary school children, there is also a need to target older school pupils in a secondary school setting that are completing their studies and moving towards independent lifestyles where they are more responsible for their life choices and behaviors. Of course, education should not just be limited to formal settings. The use of games can also serve as effective tools for education and engagement across all levels of society (Wu & Lee, 2015). HE students can also have a key role in developing and delivering these. Thus, educational institutions are integral in elevating the types of citizen climate governance needed within both the science and policy domain as part of a new civil politics of climate change that is equitable and just.

REFERENCES

Allen, M., Babiker, M., Chen, Y., Taylor, M., Tschakert Australia, P., Waisman, H., ... Waterfield, T. (2018). Drafting authors: Summary for Policymakers SPM. Retrieved from https://www.ipcc.ch/site/assets/uploads/sites/2/2019/05/SR15_SPM_version_report_LR.pdf

Bardsley, D. K., & Bardsley, A. M. (2007). A constructivist approach to climate change teaching and learning. *Geographical Research*, *45*(4), 329–339.

Corner, A., Roberts, O., Chiari, S., Völler, S., Mayrhuber, E. S., Mandl, S., & Monson, K. (2015). How do young people engage with climate change? The role of knowledge, values, message framing, and trusted communicators. *Wiley Interdisciplinary Reviews: Climate Change*, *6*(5), 523–534.

Cotton, D., & Winter, J. C. N. (2010). It's not just bits of paper and light bulbs. A review of sustainability pedagogies and their potential for use in higher education. In P. Jones, D. D. Selby, & S. R. Sterling (Eds.), *Sustainability education: Perspectives and practice across higher education* (p. 41). London: Earthscan.

Dieleman, H., & Huisingh, D. (2006). Games by which to learn and teach about sustainable development: Exploring the relevance of games and experiential learning for sustainability. *Journal of Cleaner Production*, *14*(9–11), 837–847.

Drayson, R., Bone, E., Agombar, J., & Kemp, S. (2014). *Student attitudes towards and skills for sustainable development*. York: SAGE Publications.

Gayford, C. (2002). Controversial environmental issues: A case study for the professional development of science teachers. *International Journal of Science Education*, *24*(11), 1191–1200.

Haklay, M. (2013). Citizen science and volunteered geographic information: Overview and typology of participation. In *Crowdsourcing geographic knowledge* (pp. 105–122). Dordrecht: Springer. Accessed on October 3, 2016. doi:10.1007/978-94-007-4587-27

Haklay, M. (2015). *Citizen science and policy: A European perspective*. Washington, DC: Wilson Center.

Hess, D. J., & Maki, A. (2019). Climate change belief, sustainability education, and political values: Assessing the need for higher-education curriculum reform. *Journal of Cleaner Production*, *228*, 1157–1166. Accessed on July 24, 2019. doi:10.1016/J.JCLEPRO.2019.04.291

Howarth, C., & Monasterolo, I. (2016). Understanding barriers to decision making in the UK energy-food-water nexus: The added value of interdisciplinary approaches. *Environmental Science & Policy*, *61*, 53–60. Accessed on May 28, 2018. doi:10.1016/J.ENVSCI.2016.03.014

Jasanoff, S., & Wynne, B. (1998). Science and decision-making. In S. Rayner & E. Malone (Eds.), *Human choice and climate change* (Vol. 1, pp. 1–112). Columbus, OH: Battelle Press.

Johnson, R. M., Henderson, S., Gardiner, L., Russell, R., Ward, D., Foster, S., … Eastburn, T. (2008). Lessons learned through our climate change professional development program for middle and high school teachers. *Physical Geography*, *29*(6), 500–511.

Kennedy, R. (2019). Sir David Attenborough warns of 'collapse of civilisations' at COP24 summit | Euronews. *Eoronews*. Retrieved from https://www.euronews.com/2018/12/03/sir-david-attenborough-warns-of-collapse-of-civilisations-at-cop24-summit. Accessed on November 1, 2019.

Kolb, D. (1984). *Experiential learning: Experience as the source of learning and development.* Englewood Cliffs; London: Prentice-Hall.

Kythreotis, A., Mantyka-Pringle, C., Mercer, T. G., Whitmarsh, L. E., Corner, A., Paavola, J., … Castree, N. (2019). Citizen social science for more integrative and effective climate action: A science-policy perspective. *Frontiers in Environmental Science*, 7, 10. Accessed on February 5, 2019. doi:10.3389/fenvs.2019.00010

Lawson, D. F., Stevenson, K. T., Nils Peterson, M., Carrier, S. J., Strnad, R. L., & Seekamp, E. (2019). Children can foster climate change concern among their parents. *Nature Climate Change*, 9, 458–462. Accessed on October 28, 2019. doi: 10.1038/s41558-019-0463-3

Leal Filho, W., Shiel, C., Paço, A., Mifsud, M., Ávila, L. V., Brandli, L. L., … Caeiro, S. (2019). Sustainable Development Goals and sustainability teaching at universities: Falling behind or getting ahead of the pack? *Journal of Cleaner Production*, *232*, 285–294.

McCaffrey, M. S., & Buhr, S. M. (2008). Clarifying climate confusion: Addressing systemic holes, cognitive gaps, and misconceptions through climate literacy. *Physical Geography*, *29*(6), 512–528.

Mercer, T. G., Kythreotis, A., Robinson, Z. P., Stolte, T., George, S. M., & Haywood, S. K. (2017). The use of educational game design and play in higher education to influence sustainable behaviour. *International Journal of Sustainability in Higher Education*, *18*(3), 359–384. Accessed on February 14, 2017. doi:10.1108/IJSHE-03-2015-0064

Molthan-Hill, P., Worsfold, N., Nagy, G. J., Leal Filho, W., & Mifsud, M. (2019). Climate change education for universities: A conceptual framework from an international study. *Journal of Cleaner Production*, *226*, 1092–1101. Accessed on July 24, 2019. doi:10.1016/J.JCLEPRO.2019.04.053

Narksompong, J., & Limjirakan, S. (2015). Youth participation in climate change for sustainable engagement. *Review of European, Comparative and international environmental law*, *24*(2), 171–181.

Pivec, M. (2007). Editorial: Play and learn: Potentials of game-based learning. *British Journal of Educational Technology*, *38*(3), 387–393.

Prins, G., & Rayner, S. (2007). Time to ditch kyoto. *Nature*, *449*(7165), 973–975. Accessed on November 6, 2016. doi: 10.1038/449973a

Robinson, Z. P. (2015). Are geography students good "environmental citizens?" A comparison between year of study and over time. *Journal of Geography in Higher Education*, *39*(2), 245–259.

Spicer, A. (2019). The Extinction Rebels have got their tactics badly wrong. Here's why. *The Guardian*. Retrieved from https://www.theguardian.com/commentisfree/2019/apr/19/extinction-rebellion-climate-change-protests-london. Accessed on November 1, 2019.

UNESCO. (2014). Education for sustainable development (ESD). Retrieved from www.unesco.org/new/en/education/themes/leading-the-international-agenda/education-for-sustainable-development/

United Nations. (2019). *The sustainable development goals report*. Retrieved from https://unstats.un.org/sdgs/report/2019/The-Sustainable-Development-Goals-Report-2019.pdf. Accessed on October 23, 2019.

Wu, J. S., & Lee, J. J. (2015). Climate change games as tools for education and engagement. *Nature Climate Change*, *5*(1), 413–418.

7

OCEAN OF LOVE – SCIENCE, POLICY, AND SPIRITUALITY OF THE WORLD WATER CRISIS

Marc Handley Andrus

Water as a prism for life and life-giving action in a time of climate change, overfishing, and pollution is a frame for understanding how we have come to where we are now. Water is also the lens through which we may see a way to thrive even as we face the greatest challenge of history. I believe deep scientific knowledge, comprehensive policies, and the actions of both individuals and communities all contribute to either the crisis of water, or the preservation of life, dependent on water.

However, at the heart of saving grace is seeing truly the key driver of our crisis – the water crisis (the climate crisis) is a spiritual crisis. Put another way, as people of faith, following the way of Jesus, we know that the inner world is real. There is an ocean inside of us, as well as an outer world – of water, of rivers, lakes, and the world Ocean. Jesus helped us understand the importance of the world in the heart:

> *Then he called the crowd to him and said to them, Listen and understand: it is not what goes into the mouth that defiles a person, but it is what comes out of the mouth that defiles.*
>
> (Matthew 15:10,11)

In other words, the crisis we see outside of us has its origin within us. How we meet the greatest threat the world has ever encountered begins in the human heart, with a vulnerability to love at the heart of the universe. This divine love spills over and allows us to receive anew sacred stories, new inspiration from the arts, and a capacity to connect to all of life.

In 2010 I had a personal experience of the reality described above, of how the water crisis flows forth from human choices, and then how those choices come back to us as external blessings or threats. The Diocese of California and the Diocese of Parana, Brazil, had co-hosted an Anglican climate change conference in the Dominican Republic, running at the same time as the UN climate summit in Cancun. (An active participant and signatory to the conference agreement was the new lead Primate on environment, the Most Reverend Julio Murray.)

After the conference the co-host representatives moved to a hotel on the Dominican Republic coast to debrief and plan the next steps. While there, the Rev. Dr Joshua Griffin and I hired a guide to take us snorkeling among the coral, just offshore from the hotel. I was more and more dismayed as we swam in the clear, sparkling water – there were no fish. Finally, our guide, swam down and pointed to one solitary, small fish. When we surfaced near the boat, fumes hanging just over the surface of the water made me nauseous. What was going on?

I determined to research the question when we got back to California. What I found was this:

- There are about 11,000 hotel rooms in the complex along those beaches where we stayed.
- There is no sewage treatment system for the hotels. Effluent material flows into the ocean.
- There is an oil refinery just out of sight of where we were swimming.
- The predator fish (those we eat) have been overfished in that area.

Summing up what I learned, the environmental degradation I experienced along the coast in the Dominican Republic may be seen as the externalization of human choices – the defilement begins in the human heart. I should add that the environmental degradation that I witnessed in the Dominican Republic is easily mirrored off the coast of the United States, my home. I write from no sense of a superior position.

In a global and historical context, however, the question needs to be asked – in whose hearts did the disastrous concatenation of choices leading to a degraded coastal environment in the Dominican Republic begin? Not, it seems to me, in the hearts of Dominicans. The lens must be widened to see a history of colonialization and economic dominance by the United States and Europe to see where responsibility really lies. Not Dominican hearts, or not Dominican hearts alone. Upon deep reflection I can say that my own heart is the starting place for the water desert I experienced. It may be apt to remember the notion of original sin here; I am born into a culture that is already deformed by a complex of choices that present themselves to me as a "given" part of the surrounding landscape.

When I say that the water crisis is a spiritual crisis, and follow that up by saying that what we see and experience as external to us is in fact connected to our inner lives, it is important to go further and say that the ultimate force in shaping both the inner and outer landscapes is not my own mind, nor even the minds of all people collectively, as powerful as that collective will is. Rather, it is the presence of Christ in our hearts, Christ in our world that is determinative. I am reminded of the prophecy of Simeon in the Gospel of Luke, as he beholds the infant Jesus, brought by his parents to the temple in Jerusalem:

> *Then Simeon blessed them and said to his mother Mary, "This child is destined for the falling and the rising of many in Israel, and to be a sign that will be opposed so that the inner thoughts of many will be revealed – and a sword will pierce your own soul too."*
>
> (Luke 2:34,35)

The presence of Christ is the truth that lays bare our inner thoughts; the thoughts we would be willing for the world to see and those of which we are afraid. In all that I write following, it is well to remember that while our best efforts are needed in advocacy and policymaking, in advancing and protecting scientific research in water, the most important contribution and effort of Christian bodies involved in climate and environmental work is the act of prayer, of resorting to the One who made and sustains the world.

This chapter begins with a reflection, followed by text from Genesis 6: 1–8, reminding us of holy history around water – the threat and the gift. Highlights follow about the challenge before us; that is, the status of water, and some of the ways that the part of the Anglican Communion where I serve, the Episcopal Church, is supporting care of creation.

Reflection: As a way to consider the interconnections between water, policy, and spirituality, I encourage the reader to pause for a moment to center their thoughts. With each intake of breath, consider the particularity of a wave, akin to our own individuality; and, with each exhalation, invite an awareness of connection to the vast ocean that contains all of life. Breathing in, breathing out, rest in awareness of your place in God's creation, depending on the earth, and connected to the universe.

What did this reflection bring to you? This moment of reflection at the Moana Conference was important to my keynote address in that we each need to locate the nexus of policy, spirituality, and life choices in the center of our being, in our heart, from which all flows.

THE WATER CRISIS IS A SPIRITUAL CRISIS

Let's begin by looking together at one of the Bible's great narratives of sacred history, one that centers on both the spiritual health or lack thereof of people, and on water – the story of the Great Flood.

Genesis 6:1–8

> *6 [1] When people began to multiply on the face of the ground, and daughters were born to them, [2] the sons of God saw that they were fair; and they took wives for themselves of all that they chose. [3] Then the Lord said, "My spirit shall not abide[a] in mortals forever, for they are flesh; their days shall be one hundred twenty years." [4] The Nephilim were on the earth in those days – and also afterward – when the sons of God went in to the daughters of humans, who bore children to them. These were the heroes that were of old, warriors of renown.*

> *[5] The Lord saw that the wickedness of humankind*
> *was great in the earth, and that every inclination of*
> *the thoughts of their hearts was only evil continually.*
> *[6] And the Lord was sorry that he had made*
> *humankind on the earth, and it grieved him to his*
> *heart. [7] So the Lord said, "I will blot out from the*
> *earth the human beings I have created – people*
> *together with animals and creeping things and birds*
> *of the air, for I am sorry that I have made them." [8]*
> *But Noah found favor in the sight of the Lord.*

This text is vivid in its imagery, using the flooding waters as a commentary on choices by humankind leading to destruction of all of life but that small remnant protected by God. As the waters rise in the Ocean, the waters held beneath the sluice gates of our minds strain, more and more, seeking to prevent the overwhelming of our consciousness by the truth. The truth that is pressing against our consciousness, asking to be seen and acknowledged, is one that the sacred narratives of the Bible alerted their hearers to heed. The story of the Flood is meant to awaken its hearers to the state of humanity apart from a true orientation to God. The Genesis storyteller begins the story of the Flood by telling the state of humanity from God's point of view: humanity seeks to vaunt itself, rise beyond mortal boundaries, and tend toward violence and evil.

So God takes in hand the waters of the world, a power so vast and almost limitless that humans cannot hope to control it themselves. God increases the world's Ocean by opening the floodgates of heaven, to the end that wicked humanity, and almost all life is swept away, and only a righteous remnant is left to repopulate the world. Then, God sets the rainbow in the sky as a sign that never again will God destroy the world with water.

In the sweep of the narrative and theology of the Hebrew Scripture, which contains the story of Noah and the Flood above, the import of this story is that the hearers are meant to see that they are the descendants of Noah. The responsibility to live into all that the covenant signed by the rainbow is upon all such descendants, upon us.

A second text from the Gospel of Mark reminds us of a different reality; that is, that it is Jesus Christ who sustains God's promise to protect the world from all that the storm, the wind and the waves, represent.

Mark 4: 35–41

> 35 *That day when evening came, he said to his*
> *disciples, "Let us go over to the other side."*
> 36 *Leaving the crowd behind, they took him along,*
> *just as he was, in the boat. There were also other*
> *boats with him.* 37 *A furious squall came up, and the*
> *waves broke over the boat, so that it was nearly*
> *swamped.* 38 *Jesus was in the stern, sleeping on a*
> *cushion. The disciples woke him and said to him,*
> *"Teacher, don't you care if we drown?"* 39 *He got up,*
> *rebuked the wind and said to the waves, "Quiet! Be*
> *still!" Then the wind died down and it was*
> *completely calm.* 40 *He said to his disciples, "Why are*
> *you so afraid? Do you still have no faith?"* 41 *They*
> *were terrified and asked each other, "Who is this?*
> *Even the wind and the waves obey him!"*

The story above asks us to consider ourselves, to see if we live in the light of a covenant God made with humanity – to be people of peace, loyal to God and the vision of a world community. Or is it the case, the story asks, that we have become again like the humanity that preceded the Flood? This question – "Who are we now?" – is the one that strains the sublimating gates of our consciousness.

THE ANTHROPOCENE: SCIENCE EXPRESSES THE THEOLOGICAL INSIGHT OF OUR BECOMING OUR OWN GODS

For us, the Genesis story of the Flood is real in its entirety within our hearts, the core of ourselves where we work out who we are and what we value, where we reconcile our mistakes through forgiveness, where we meet God. The events of the Flood are real within that most real of places, our hearts. The suffering already being experienced by millions due to the near horizon of climate change, and the far greater climate chaos that is likely to come, while it begins in our hearts and minds, has an exterior, experienced reality.

The Flood is a sacred story that awakens its listeners to the negative and the loving possibilities of life. The Flood, as sacred story, is the act of a loving God. The consequences of the Anthropocene are the unconscious, heedless acts of collective humanity, a pseudo-god.

Like the women and men who wrote the sacred texts that became our Bible, there are storytellers today seeking to help us understand our inner and outer worlds. There is an abundance of contemporary films and novels all pointing to a single truth: humanity has taken the position of a god, and we now influence the precious balance of the atmosphere, the quality of even the vast world ocean, and the health of the land. These modern-day storytellers are also united on another point – humans are ill-equipped to be gods. Unintended consequences rush back toward people, and all of life, the evil fruits of our worst inner thoughts, of greed, desire for dominance and power.

While God, in the covenant made with humanity through Noah, promised to never again destroy the Earth by means of flooding waters, we, as a collective pseudo-god have made no such promise to ourselves or to life on the Earth. Many of the apocalyptic and post-apocalyptic novels and films envision climate chaos that centers on monstrous floods, inundating coastal cities, and overwhelming island communities. Here are a few of the films and novels that make the point about our rampaging inner lives manifesting in an outer experience of climate chaos: *The Day after Tomorrow* (2004), *Geostorm* (2017), *Aquaman* (2018), *Trail of Lightning* (2018), and *Storm of Locusts* (2019). The foregoing is only a small sample of such post-apocalyptic stories.

ANTHROPOCENE OUTCOMES

Geologists identify eras and epochs of the Earth's history by prominent elements that appear in the rock and soil strata. We have been living in the Holocene Epoch, which began just over 10,000 years ago. "Holocene" means "completely new," and refers to the appearance in the stratigraphy of the Earth of an efflorescence of human artifacts, as well as the appearance of urban centers.

The consensus of the geological science community is that we are entering a new epoch, which they have named the Anthropocene, meaning that for the first time in the life of the planet, the determinative factor in shaping the climate (and the climate-related deposits that can be read by geologists) is human activity. Science is confirming what the storytellers of the distant past and of today have told us – humans have become as gods.

Outcomes from unsustainable choices related to water can be described by conditions of both water quality and water quantity; that is, either too much or too little water. Among the impacts are a growing plastic buildup in the ocean, species loss, rising sea waters, and climate refugees, among others.

PLASTIC ISLAND, PERVASIVENESS OF PLASTIC WASTES

The Great Pacific Garbage Patch

Researchers from The Ocean Cleanup project claimed that the Great Pacific garbage patch covers 1.6 million square kilometers. The plastic concentration is estimated to be up to 100 kilograms per square kilometer in the center, going down to 10 kilograms per square kilometer in the outer parts of the patch. An estimated 80,000 metric tons of plastic inhabit the patch, totaling 1.8 trillion pieces. Ninety-two percent of the mass in the patch comes from objects larger than 0.5 centimeters, while 94% of the total objects are represented by microplastics.

The North Atlantic garbage patch is an area of man-made marine debris found floating within the North Atlantic Gyre, originally documented in 1972. The patch is estimated to be hundreds of kilometers across in size,[2] with a density of over 200,000 pieces of debris per square kilometer (one piece per 5 square meters, on average). The debris zone shifts by as much as 1,600 km (990 mi) north and south seasonally, and drifts even farther south during the El Niño-Southern Oscillation, according to the National Ocean Atmospheric and Administration(NOAA).

In 1950, the world's population of 2.5 billion produced 1.5 million tons of plastic; in 2016, a global population of more than 7 billion people produced over 320 million tons of plastic. This is set to double by 2034. Every day approximately 8 million pieces of plastic pollution find their way into our oceans. There now may be around 5.25 trillion macro and micro-plastic pieces floating in the open ocean, weighing up to 269,000 tons.

SPECIES LOSS AND ENDANGERMENT

Large pelagic sharks, including threshers, great whites, and hammerheads have declined more than 75% in the Northwest Atlantic since 1986. Similar reductions over the past 50 years have been reported for other predatory marine fishes. One of these, the Atlantic cod, a species that supported one of the largest and economically most important fisheries throughout the North Atlantic for almost 500 years, has declined by as much as 99.9% in some areas since the early 1960s. Failure to arrest these declines will result in the loss of species from parts of their current geographical ranges; a fate already experienced by more than 55 species.

Many of us have personal experiences of species loss, including species loss in aquatic systems, due to climate change, pollution, and/or overfishing.

ALASKA TRIBAL GROUPS AND SALMON

In another different ecosystem – one in the Arctic – Native Americans in Alaska have been in an interlocking relationship with salmon in the Alaskan rivers for 10,000 years.

> *The salmon are vanishing. And it is not only the number of fish in decline but also their size. I was shown old photographs of king salmon the same length as the people who caught them, 36kg monsters that were common. Now a good-sized fish is 9kg, and smaller fish lay fewer eggs. Whilst roughly five salmon used to return for every spawning adult, these days it is not far off one-to-one. Before 1997, the historic average for the king salmon run on the Yukon river was 300,000 fish;*

> *in 2013, just 37,000 came back. All this is playing out against the background of a rapidly changing climate, with the Arctic warming twice as fast as the global average. The river's winter ice is unreliable. As traditional sources of food disappear, people are forced into buying from the stores – where the food has often been flown thousands of miles and has a price tag to match. Their villages are often many hundreds of miles from the nearest road.*
>
> *As local diets change, instances of once unknown illnesses such as heart disease and diabetes are also escalating. Forced into a cash economy, but with few jobs available, there is a widening gap between rich and poor, a growing dependence on food stamps and government handouts. Aimless and rootless, many people are turning to drink. Alcoholism and its associated symptoms – crime, domestic violence, suicide – are taking a heavy toll. As the culture erodes, so does the language, and a sense of pride in the unique way of life of some of the planet's last remaining hunter-gatherers.*
>
> (Weymouth, 2018a, 2018b)

I quoted the above piece by Adam Weymouth at some length because in his in-depth reporting (the reporter actually took several months canoeing along the Yukon and other Alaskan rivers to make his report) takes us intimately into the terrible loss through an entire ecosystem.

WELSH TRADITIONAL RIVER FISHING

Far from the world of the Gwich'in, in Wales, comes another story of an indigenous population in a centuries-old relationship with other species, and a new time of crisis. On our way to a conference in Llandovery, Wales, my wife, Sheila and I transferred from the train to a bus for the last leg of the journey. With a three-hour wait for the next bus, we wandered down to the annual Carmarthen River Festival.

At the festival, we met a group of Welsh traditional fisherfolk, who fish the Carmarthen River using coracles, small round boats that have been used in Wales for over 2,000 years. The fish that the local fishing industry relies upon are ocean trout and salmon. In the United Kingdom, there are 2018 stock assessments for salmon and trout that indicate that stocks are now performing at their worst levels on record. All salmon stocks and about half of sea trout stocks are below safe biological levels and at risk of ongoing decline. The new assessments are of great concern and indicate a deeply worrying trend of ongoing decline. The declines in ocean trout and salmon stocks in Wales are the result of industrial fishing in the ocean, where giant netting operations take in all the sea life in the scope of the nets, regardless of species. The end result is the drastic depletion of these fish species, and the probable end of a traditional way of life that has endured and been sustained for many centuries.

RISING OCEAN LEVELS

Global sea levels could rise by more than 2 meters by the end of this century if emissions continue in a "business as usual" scenario, recent research has warned. The international team of researchers said this would have "catastrophic" consequences for the whole world, forcing 187 million people off their land (Bamber, Oppenheimer, Kopp, Aspinall, & Cooke, 2019). Parts of London, Los Angeles, New York and Rio de Janeiro could be submerged, leaving 1.79 million km^2 of land lost globally, in a significant upwards revision of Intergovernmental Panel on Climate Change sea level estimates, according to the study. "The results were worse than we anticipated," lead researcher Jonathan Bamber, from the University of Bristol told The Independent. If emissions continue to grow unchecked in the "business as usual" scenario, global temperatures could increase by 5 °C by 2100, the report states.

Long before the Pacific Ocean subsumes thousands of low-lying islands, waves will begin washing over them frequently enough to ruin groundwater supplies and damage crops and fragile infrastructure. Reports say this likely will render many coral atolls uninhabitable within decades – before 2030 in a worst-case scenario and by 2065 in a more optimistic one.

"When you're walking around the islands and you see kids there, you realize this isn't something that's next generation or the generation after that," said Curt Storlazzi, a scientist with the US Geological Survey and the study's lead author. "If these scenarios play out, we're talking about those kids' generation." (Storlazzi et al., 2018).

CLIMATE REFUGEES

Currently, forecasts vary from 25 million to 1 billion environmental migrants by 2050, moving either within their countries or across borders, on a permanent or temporary basis, with 200 million being the most widely cited estimate, according to a 2015 study carried out by the Institute for Environment and Human Security of the United Nations University (Hilft, 2015). The majority of refugees today come from five countries: Syria, Afghanistan, South Sudan, Myanmar, and Somalia. War and unrest cause these refugees to flee.

According to Eurostat, EU member states received over 1.2 million first-time asylum applications in 2015: more than double that of the previous year. Four states (Germany, Hungary, Sweden, and Austria) received around two-thirds of the EU's asylum applications in 2015, with Hungary, Sweden, and Austria being the top recipients of asylum applications per capita. More than 1 million migrants crossed the Mediterranean Sea in 2015, dropping considerably to 364,000 in 2016. Numbers of arriving migrants fell again in 2017.

Focusing on Syria alone, there are in total approximately 5 million externally displaced persons.

The number of family units – usually mothers or fathers with small children – apprehended in the US Customs and Border Protection's Rio Grande Valley sector jumped from 49,896 in fiscal year 2017 to 63,278 in fiscal year 2018, which ended September 30. That's a 27% increase, according to recently released agency statistics. According to the UN High Commissioner for Refugees, in 2017, 65.6 million people were forcibly displaced worldwide because of persecution, conflict, violence, or human rights violations alone.

FRESH WATER SCARCITY

Somewhat over 2 billion people lack access to clean drinking water and 4.2 billion people lack access to adequate water sanitation facilities (WHO/UNICEF, 2019) Without significant changes, by 2025 two-thirds of the world's population will face water scarcity. The global population tripled in the twentieth century, but water consumption increased sevenfold. By 2050, after we add another billion people to the population, water needs will rise 80% above current demands.

The term "water apartheid" has been used to describe the current situation in the Global South. The world's poor are living without water in areas that do not have enough water to begin with (Africa), where surface water has become severely polluted (S. America, India), or both (Northern China). Most of the world's megacities – those with 10 million or more inhabitants – lie within regions experiencing water stress.

The UN estimates that by 2030 more than half of the largest cities will have slum dwellers with no access to water or sanitation services whatsoever. One report cited a current example of an area in Mumbai, where one toilet serves 5,440 people (UN-Habitat, 2004).

An average person needs 50 liters/day for drinking, cooking, and sanitation. The average North American uses almost 600 liters/day. The average African uses six. Using the lens of water in this time of the Anthropocene, a key question is: how should we, in the global human community, respond to the Anthropocene?

FAITH MEETS THE ANTHROPOCENE: INTERCONNECTION OF LIFE

In considering the impacts of the Anthropocene and how our unsustainable life choices spread across ecosystems for ill, I'm reminded of words from Martin Luther King Jr in his Letter from a Birmingham Jail:

> *In a real sense all life is inter-related. All men are caught in an inescapable network of mutuality, tied in a single garment of destiny. Whatever affects one directly, affects all indirectly. I can never be what I ought to be until you are what you ought to be, and you can never be what you ought to be until I am what I ought to be...This is the inter-related structure of reality.*
>
> (King & Martin Luther King, Jr. Center for Nonviolent Social Change, 1968)

This is the larger reality of the Anthropocene – we are interconnected for good and for ill, and the solutions we choose will most certainly reflect this reality.

THE ROLE OF FAITH BODIES IN ADDRESSING THE WATER CRISIS

Where, in all of the bad news, can people of faith find hope and inspire others? The answer in part comes from our sacred stories and the faith we carry by God's grace, understanding that with God's help we can do more than we could ever hope or imagine. For action, the "we" is key. Five billion people on planet earth today – the majority of the world's population of approximately 7 billion people – are people who claim a faith tradition. And, the world's faith traditions hold the Earth in reverence – a sacred gift of which we are a part. Many religions have beliefs and rituals of transformation, seeking conditions of harmony, integrity, and peace. There are countless people of faith already working to protect and heal the environment.

Ecumenical and interfaith environment and climate action alliances have formed with relative ease around a common mission. Our own former Archbishop of Canterbury, the Rt. Rev. Rowan Williams, reminded the communion that "It takes a global body to address a global problem." Our global relationships can empower our action and advocacy across the globe.

Nongovernmental organizations (NGOs) numbering over 100,000 worldwide are working on various elements of the climate/environment challenge, including water. The author Paul Hawken notes in *Blessed Unrest* that this collection of NGOs is largely not networked, but he hopefully asserts that the network functions like the earth's immune system. Granting the truth of Hawken's metaphor, could we who are people of faith think of connecting our efforts as a kind of immune system booster, a shot of world vitamins from the faith communities?

THEOLOGY IN ACTION: THE POLICIES OF THE EPISCOPAL CHURCH IN RESPONSE TO THE ENVIRONMENTAL CRISIS

In July 2018, the Episcopal Church gathered for its triennial General Convention, wherein program, budget, and priorities are voted upon by representatives from every diocese and all bishops of our province. At the 2018 General Convention, a historic number of environmental resolutions were approved related to care of creation. These environmental justice and climate-change related resolutions reflect not only the commitments of Episcopalians that have been growing and deepening over decades. Also, reviewing the body of environmental resolutions from the 2018 General Convention, I believe they reflect a shared awareness of the reality of interconnection – that is, that in solving environmental problems, we need also to be aware of the connection of the socioeconomics and health (of people and ecosystems) connected to those problems. The areas addressed by these policies include support for a just economic transition for those most affected by climate change, tree planting, carbon offsets, support for Paris Agreement goals, fossil fuel financial divestments, water as a human right, protection of ocean health, use of tools such as the carbon tracker sustainislandhome.org to support transformation of habits, among others.

The environmental policy resolutions of the Episcopal Church reflect the many facets of water and climate change and environmental degradation. No one group within the Episcopal Church will in likelihood address every part of each of these environmental resolutions. Nor will it be the Anglican Communion alone that can accomplish the vast work needed to face into environmental injustice and climate change. Surely, we must act together as people of faith across the earth. And, we must make good partnerships with all the other sectors of society – business, city, state, regional, and national governments, educational institutions, health-care institutions, and tribes, to name some of the constituent sectors represented in the We Are Still In movement in the United States.

OCEAN OF LOVE

As I opened with a centering exercise, I'd like to invite you, the reader to close with another spiritual practice. There is a mudra, a symbolic hand gesture, in Buddhism called Ocean. It is a simple gesture – the palms of the hands opened and turned toward one another. The mudra is called "Ocean" because its meditative meaning is that between the palms of your hands is contained the fullness of your being in all its integrated complexity. I invite you to prayerfully make this gesture, the Ocean mudra, and as you quietly and slowly breathe watch the thoughts that arise when you invoke the image of the fullness of your being. Memories of people who have influenced you may come to mind; determinative experiences, both painful and joyful; qualities of yourself that you regard as strengths or as weaknesses. Witness all this with attentiveness and receptivity. Then, finally, look to the presence of the living Christ in the center. Once seen, how does the presence of Christ interact, change, integrate all the oceanic elements that have arisen in this meditation? End by asking Christ to bless and calm the ocean of yourself, that you might be a greater blessing to the world around you.

REFERENCES

Bamber, J. L., Oppenheimer, M., Kopp, R., Aspinall, W., & Cooke, R. (2019). Ice sheet contributions to future sea-level rise from structured expert judgement. *Proceedings of the National Academy of Sciences of the United States of America*, *116*.

Hawken, P. (2007). *Blessed unrest: How the largest social movement in history is restoring grace, justice, and beauty to the world*. Penguin Books.

Hilft, B. E. (2015). *World risk report*. United Nations University—Institute for Environment and Human Security (UNU-EHS), Bonn.

King, M. L., & Martin Luther King, Jr. Center for Nonviolent Social Change. (1968). Letter from a Birmingham jail.

OR&R's Marine Debris Program. Garbage patches. Retrieved from https://marinedebris.noaa.gov/info/patch.html

Storlazzi, C. D., Gingerich, S. B., van Dongeren, A., Cheriton, O. M., Swarzenski, P. W., Quataert, E., … McCall, R. (2018). Most atolls will be uninhabitable by the mid-21st century because of sea-level rise exacerbating wave-driven flooding. *Science Advances*, *4*(4), 1–9, eaap9741.

The Ocean Cleanup. Retrieved from https://theoceancleanup.com

UN-Habitat. (2004). The challenge of slums: Global report on human settlements 2003. *Management of Environmental Quality: An International Journal*, *15*(3), 337–338.

UNHCR. Retrieved from https://www.unhcr.org/uk/

Weymouth, A. (2018a, June 19). The decline of wild salmon in Alaska's Yukon us putting lives and native culture at risk. Chalk Valley History Festival History Hub. Retrieved from https://cvhf.org.uk/history-hub/the-decline-of-wild-salmon-in-alaskas-yukon-is-putting-lives-and-native-culture-at-risk/. Accessed on March 13, 2020.

WeymouthA. (2018b). The decline of wild salmon in Alaska's Yukon is putting lives and native culture at risk. Retrieved from https://inews.co.uk/news/long-reads/alaska-wild-salmon-kings-of-the-yukon-adam-weymouth-540173

WHO. Water supply, sanitation and hygiene monitoring. Retrieved from https://www.who.int/water_sanitation_health/monitoring/coverage/en/ quoted on https://www.un.org/en/sections/issues-depth/water/. Accessed on March 6, 2020.

8

READING THE BIBLE AS WATERS RISE: ECOLOGICAL INTERPRETATION OF SCRIPTURE

Emily Colgan

> *The greatest failure of Christianity in the total course of its history is its inability to deal with the devastation of the planet.*
>
> (Thomas Berry, quoted in Collins, 1995, p. 152)

In the months since this chapter began taking shape humanity has witnessed an alarming increase in extreme weather events. We have seen a 'record-breaking' super typhoon in Japan (Mori, 2019), 'megafires' in the Australian bush country (Foster, 2019), 'devastating floods' in India (Burdon-Manley, 2019) and a 'one-in-a-thousand year' heatwave in Europe (Watts, 2019). By the end of July, the melting of Greenland's ice sheets had reached levels that climate models had projected for 2070 'in the most pessimistic scenario' (McFall-Johnsen, 2019). The force of Hurricane Dorian, the 'most powerful storm to hit the Bahamas'

('Hurricane Dorian: Path of Destruction', 2019), led some to claim that the Category 1–5 Saffir–Simpson scale for rating hurricanes is no longer adequate for measuring the intensity of such extreme storms (Masters, 2019). The list of these weather events could (and does) go on. And on. This desperate situation has prompted some Christians to begin wrestling with what it might mean to construct a faith-based response to the climate crisis. Part of this response relates to the interpretation of Scripture. This chapter outlines a strategy for reading the Bible as Christian communities explore how this text might speak into the contemporary context of climate crisis. Before we proceed, however, we need to take a few steps back to examine the work of a scientist named Lynn White Jr., whose 1967 seminal paper entitled, 'The Historical Roots of Our Ecological Crisis' raised some serious questions about the relationship between Christianity and ecological degradation.

In this short article, White (1967) made a compelling case arguing that a causal relationship exists between Christianity (grounded in the Bible and its history of interpretation) and the contemporary ecological crisis. Among the accusations levelled against 'Western' Christianity in this chapter is the charge that it is 'the most anthropocentric religion the world has ever seen' (p. 1205). Grounded in the creation story of Genesis 1, Christianity, argues White, gives privilege to the human community as the pinnacle of the created order, thereby diminishing the value of the other-than-human community:

> ... *God blessed them, and God said to them, 'Be fruitful and multiply, and fill the earth and subdue it; and have dominion over the fish of the sea and over the birds of the air and over every living thing that moves upon the earth'.*
>
> (Gen 1:26–28)

Not only does this ecological segregation establish humankind as 'master' over the environment, but it also depicts as God's will the exploitation of 'nature' to serve human interests (p. 1205). White suggests that it is this anthropocentric worldview that has enabled the emergence of scientific and technological ideologies in the West, which have had a devastating impact on the environment (p. 1206). And so, he argues, Christianity 'bears a huge burden of guilt' for the kind of behaviour in the world that has been so ecologically disastrous (p. 1206).[1] The ecological crisis is a religious crisis.

But White doesn't stop there. In concluding his article, White insists that since the roots of the ecological crisis are largely religious, the 'remedy must also be essentially religious, whether we call it that or not' (p. 1207). If, as White suggests, it is through the Bible (via Christianity) that human beings have become alienated from the environment, then this same source may also provide the motivation for the transformation of human attitudes towards our planet.[2] White's charge challenges biblical interpreters to critically re-examine the Bible, taking into account the severity of the ecological crisis as well as the complicity of humanity in the destruction of the

1 I appreciate that this summary – and indeed the article itself – is reductive and somewhat simplistic. It is not my intention to critically analyse the cogency of White's work here; rather, I wish to acknowledge the importance of this article in its forcing biblical scholars to re-examine the biblical depiction of humanity and the environment.

2 Writing decades later, Oelschlaeger (1994) echoes this sentiment arguing that, 'There are no solutions for the systemic causes of eco-crisis, at least in democratic societies, apart from religious narrative' (p. 5). In a similar vein, Rosemary Radford Ruether (1992) insists that the vast majority of the 2.2 billion Christians in the world will only be persuaded of an ecological consciousness if they see that it grows from the soil in which they are planted (p. 207).

Earth. It exhorts readers to re-examine the anthropocentric assumptions that characterise (parts of) the biblical text and its long history of interpretation, and urges these readers to reflect critically upon the role of these texts in the social construction of reality.[3] In short, White's assertions challenge interpreters to read the Bible in such a way that it might contribute to the emergence of an ecological sensibility that is appropriate to the environmental concerns of the twenty-first century.

Unsurprisingly, White's article proved provocative in the area of biblical interpretation and the responses to this paper have been many and varied.[4] It is not within the scope of this chapter to include yet another analysis of this work. Rather, I wish to pick up on White's idea that 'more science and more technology are not going to get us out of the present ecologic crisis until we find a new religion, or rethink our old one' (p. 1206). White seems to be suggesting that Christian stories about who we are as humans – our theological anthropologies – are no longer appropriate in the contemporary context. He is arguing that we need a new story – or new stories – about our relationship to the world around us. And I think he might have a point.

Because fast forward 50 years and scientific leaders continue to warn us that climate change is real, and it is already causing irrevocable damage to the planet. Where questions of human complicity once dominated the climate change discourse, scientific consensus now holds humanity responsible, making links between climate change and human initiatives such as deforestation, fossil fuel emissions, pollution, overpopulation, the destruction of ecosystems and the

3 Cf. Schüssler Fiorenza (1999, p. 11).

4 Scharper (1999) offers a useful summary of the initial responses to White's article.

reduction of biodiversity. Debate no longer centres primarily around whether or not climate change exists, or even whether humanity can avert it; rather, the dialogue has become one of mitigating the potentially catastrophic consequences of this global shift.

Yet despite climate change being acknowledged as among the most critical issues faced by the Earth community, this unprecedented threat to the planet has failed to galvanise humanity, and questions remain as to whether the political will exists to address the situation quickly and effectively.[5] Scientific fact alone, it seems, is not enough to counter what Thomas Berry (1996) calls the 'deep cultural pathology' that enables humanity to cause such damage without thought of long-term consequence (p. 410). For these warnings to challenge the well-established patterns of human ecology, it seems they must also find a degree of coherence with the religious and philosophical foundations that underlie the way in which human communities perceive themselves and their relationship to the surrounding environment. And it is this context which has led to the emergence of ecological readings of the Bible.

ECOLOGICAL INTERPRETATION

The reading strategy outlined below was developed by the Earth Bible Team and comes out of South Australia.[6] It

5 Darragh (2000) notes that environmental policy almost always takes third place after the undisputed first place of economics and the second place of social policy (p. 162).

6 This approach was developed initially in a number of publications including, Habel (2000b, 2001b); Habel and Wurst (2000, 2001); Balabanski and Habel (2002); and Habel and Trudinger (2008). An Earth Bible Commentary series is also in the process of publication.

provides *one* way of reading the Bible in the context of ecological degradation.[7] A defining characteristic of this approach to the biblical text is that of perspective. This reading strategy requires a radical reorientation to the biblical text whereby Earth is placed at the centre of the interpretation and questions asked of its presence there.[8] To approach the text in this way requires that the reader recognise the Earth as a subject to which the reader seeks to relate empathetically. Earth is not a theme for analysis, Earth is autonomous; it is an agent in its own right, a subject or character in the text. This approach emphasises reflection *with* Earth and attempts to perceive reality from Earth's perspective (Habel, 2000a, p. 33). Interpreters seek to give voice to the Earth's experience, ascertaining whether or not it is oppressed, silenced or liberated at any given point within the biblical narrative (Habel, 2008, p. 1). And so, with Earth as the interpretive focus, this approach attempts to explore the ways in which the biblical text might inform the current ecological crisis and the ever-increasing effects of climate change (Habel, 2009, p. 56).

To privilege the Earth's perspective in the Bible is not a neutral act. It is to give value to relations of power within the

---

7 In response to the deepening ecological crisis – and to White's accusations of Christian complicity in this predicament – a number of new hermeneutical approaches have emerged within the discipline of biblical studies. Each approach to biblical interpretation varies politically, ethically and theologically, with its own assumptions and initiatives for responding to this crisis. Horrell, Hunt, and Southgate (2008) provide a useful overview of these hermeneutical approaches.

8 A note about terminology. I will be referring to 'Earth' in a general sense, but I want to acknowledge the risk of homogenising this vast and complex phenomenon. So, while it will not be possible to articulate the incredible diversity of this realm, I am using 'Earth' to refer to the other-than-human: mountains, air, trees, lakes, animals, land, birds and, of course, moana.

biblical text and to question the priorities and hierarchies implicit within its narrative. Right from the outset, then, this interpretation is not value-neutral or 'objective' but establishes a moral stance that is at the centre of the investigation. In reading the biblical text, interpreters take up the cause of justice for Earth. This stance is informed by the six ecojustice principles outlined by the Earth Bible Team (Habel, 2008, p. 2):

1. The principle of intrinsic worth: The universe, Earth and all its components have intrinsic worth/value.[9]
2. The principle of interconnectedness: Earth is a community of interconnected living things that are mutually dependent on each other for life and survival.[10]
3. The principle of voice: Earth is a subject capable of raising its voice in celebration and against injustice.[11]
4. The principle of purpose: The universe, Earth and all its components are part of a dynamic cosmic design within which each piece has a place in the overall goal of that design.[12]
5. The principle of mutual custodianship: Earth is a balanced and diverse domain where responsible custodians can

---

9 See, for example, Gen 1: 4, 10, 12, 18, 21, 25. It is important to note, however, that intrinsic worth does not necessarily correspond to intrinsically 'good' in a moral sense. Code (2006) describes ecosystems as being as 'cruel as they are kind, as unpredictable and overwhelming as they are orderly and nurturant, as unsentimentally destructive of their less viable members as they are cooperative and mutually sustaining' (p. 6).

10 See, for example, Jer 31:35–37.

11 See, for example, Jer 50:46 and Matt 28:2.

12 See, for example, Job 38–40.

function as partners with, rather than rulers over, Earth to sustain its balance and a diverse Earth community.[13]

6. The principle of resistance: Earth and its components not only suffer from human injustices but actively resist them in the struggle for justice.[14]

These principles underline this interpretive approach to reading biblical texts, providing a moral centre against which such texts will be tested and held accountable.[15]

Bearing these principles in mind, an ecological hermeneutic then uses a three-step process to interpret a biblical text: suspicion, identification and retrieval. The degree to which anthropocentrism is embedded within most Western societies requires an ecological interpretation to begin from a position of suspicion. Any textual analysis, therefore, begins with the suspicion that a text is inherently anthropocentric and that its history of interpretation will reflect a similar bias (Habel, 2008, p. 4). That is, most biblical texts will reflect the unique interests of humanity as the centre of all purposeful activity and may serve as a potential source of ecological degradation. Similarly, most Western interpreters have inherited a strongly anthropocentric approach to reading the Bible. In the instances where the Earth is attributed agency in the biblical text, for example, such activity is typically

13 See, for example, Gen 2:15. This principle is designed to reflect the role of human beings in the Earth community, but it is not about 'stewardship'. An ecological perspective would argue that the stewardship model retains a hierarchy of power which is problematic as it assumes that humans are creatures of a different order and destined by God to rule over the Earth for God.

14 See, for example, Lev 18:28 and Jer 4:28.

15 These ecojustice principles are deliberately expressed in non-theological and non-biblical terms so as to facilitate dialogue with those beyond the Christian tradition (Eaton, 2000, p. 38).

dismissed as literary convention. From this position of suspicion, ecological interpreters begin to examine the biblical text, asking questions that are informed by the ecojustice principles. In this way readers critically examine the presentations of Earth within the biblical text, analysing the interaction between Earth and the text's other characters.[16]

Critical attention to the Earth's presence within a biblical text requires both sensitivity and an empathetic imagination. This is the task of identification, and it is primarily concerned with the orientation of the reader in approaching a dialogue with the biblical text. Norman Habel (2008) describes this step as coming to terms with one's 'deep ecological connections' (p. 3), whereby an interpreter is able to reposition him or herself as self-consciously part of the Earth community. Read from the perspective of an interconnected member of this community, a new level of sensitivity emerges. Such sensitivity encourages particular attention to the presence of Earth and compels a response to injustices perpetrated against this subject as they appear in the text. From a position of identification with the Earth, the interpreter is implored to take sides and stand in solidarity with this subject, binding him or herself to the Earth community and joining the struggle for justice. In taking up the cause of the Earth, interpretation involves being open to this Other, representing, advocating for and supporting the

16 Even with these ecocentric intentions, however, the task of sensing the Earth within a text can be problematic, as access to this perspective is disclosed through anthropocentric media. In the sense that all biblical texts have been shaped by human beings, an inescapable degree of anthropocentrism will inevitably bind a text to its human origins.

presence of this character within the biblical text. It involves imagining one's way into the reality of an/Other who is differently situated and ultimately articulating this experience.[17]

The final task of retrieval acknowledges that the Earth is already speaking and acting within a text. It is about reclaiming the narrative space of Earth. Thus, this task seeks to listen to and learn from the silenced or ignored voice of Earth, recovering the 'underside' of the biblical narrative.[18] In allowing space for expression, the Earth becomes both narrator and interpreter of the biblical text, a character central to the narrative drama and indispensable to its unfolding (Habel, 2009, p. 59). By paying particular attention to the Earth's agency, the task of retrieval additionally seeks to identify instances where the Earth demonstrates resistance to anthropocentric perspectives that exclude or oppress its presence in the text. At times this task entails entering into the silence of a text where this presence is suppressed and looking for subtle interruptions of the master narrative. Having identified the Earth's presence within a text, the interpretive task thus becomes one of giving voice to the experience of this character.

---

17 Speaking on behalf of an/Other potentially risks the presumption of assuming to know the thoughts, feelings and aspirations of that Other. In this sense, advocacy can be dangerous in its potential to identify the interests of the interpreter, rather than those of the Earth. In an attempt to avoid such pitfalls, this reading strategy recognises the partial and inadequate nature of identification. Guided by the ecojustice principles, this advocacy will not claim to fully understand or *be* the Earth; rather it will seek to listen to, see, or speak *with* and in support of the Earth.

18 Habel (2008) stipulates that this 'voice' need not correspond to the language of words associated with the human voice; rather, this term refers to a multiplicity of communicative forms including gesture, sign, image and sound that convey a message (p. 5).

In demonstrating a relatively simple ecological interpretation of a biblical text, I want to return to the idea that traditional theological anthropologies no longer resonate in the context of climate crisis and that we need new stories of who we are. More specifically, I want to return to Genesis 1 which, White argued, has played such an important role in the Christian narrative about the relationship between humanity and the rest of the Earth community.

GENESIS 1

Earth is one of the first characters introduced in the Bible as the initial verses of Genesis depict Earth's birth and God's celebration of this birth (Gen 1:9–10).[19] In this opening text, initially at least, Earth is an agent, a partner in the creation process. Earth is presented as the source of all living things: Earth's soil brings forth all kinds of vegetation (vv 11, 12); animals emerge from the Land (v 24); and Moana/Water brings forth life in the oceans (v 20). There is an explicit bond between Earth and the other-than-human community. All creatures have a common kinship that derives from a common ancestor: Earth. And God sees that this is good (vv 4, 10, 12, 18, 21, 25); God affirms, values and celebrates the birth of Earth and the other-than-human community.

With the creation of human beings in vv 26–28, however, the narrative shifts and Earth fades into the background, signalling the emergence of beings of a different order. Humans are not connected to the existing created order; they do not emerge from Earth like the other living creatures. They

19 Habel (2011) argues that Earth is distinct from all other components of the cosmos. 'Only *Erets* is revealed from below, born out of the waters; *Erets* is a hidden mystery made manifest, a sacred domain in the cosmos' (p. 32).

are not part of Earth in any way. Unlike other creatures, human *beings* are made in the image (*tselem*) of God (v 26). There is no escaping the overt hierarchy asserted by this text, writes Brett (2000, p. 77). Like God in the world, humans are given the mandate to dominate (*rada*) and subdue (*kabash*) the Earth (v 28). They are set apart from other beings and seem to have a distinctive and superior relationship with God.[20] Human beings are given priority in relation to God, to other living creatures and to Earth itself (Habel, 2008, p. 6).

The language that depicts the relationship between humans and the Earth community in this text denotes violence. Habel (2006, p. 39) argues strongly that the Hebrew term *rada* involves the forceful exercise of power.[21] It is what kings and taskmasters do; they dominate their subjects and their enemies (see I Kings 4:24; 5:16). Similarly, the Hebrew term *kabash* (subdue) is a harsh verb that reflects the use of violence. It is used elsewhere to mean 'crush under foot' (Mic 7:19), to 'subjugate' (Josh 18:1; Jer 34:11) and to rape a woman (Esth 7:9; Neh 5:5). These are actions that explicitly silence and suppress Earth and the Earth community. As it stands, these verses seem to devalue Earth and its other-than-human inhabitants, justifying the use of human violence and subjugation. There is a profound difference between humans and

---

20 Habel (2009, pp. 3–4) explores the ways in which this text has been used to foster the idea of humans as representatives of the divine ruler, exercising power over the Earth.

21 Habel (2001a) identifies three main readings of God's mandate for humans to dominate and subdue the Earth. Some see it as a hierarchal mandate that implies 'domination and oppression'; some claim that the mandate gives humans the right to be responsible stewards; and others 'soften the import' of the words, suggesting they refer to kindness and caring (p. 180). While the term *rada* might not appear exclusively in contexts of violence throughout the Hebrew Bible, its use here, in close connection to the verb *kabash* (to subdue), suggests violent overtones.

the rest of creation here. This story of who we are is disconnected from the story of plants and animals and oceans. It is removed from the grand context of the cosmos, the planets, the stars and space.

Traditional interpretations of Genesis 1 have also highlighted the unique place of humanity in this text and our superiority to the other-than-human community. The assumption derived from this text is that human domination of the other-than-human is natural and obvious. But the consequences that follow from such a position are unjust and inconsistent with the interconnected ecological reality of our planet (Habel, 2008, p. 7). This is a text that has the potential to be ecologically destructive, serving as the basis for exploitation, oppression and abuse of the Earth community. And so, the question arises as to whether there is an alternative narrative, which might contribute to the creation of a new story?

GENESIS 2

Surprisingly perhaps, we need look no further than the very next chapter of Genesis. Although often overlooked, Earth is utterly central to this second creation narrative. As the text opens, the reader learns that there is no vegetation on Earth because there is no one to 'serve' *Adamah* (Land/Ground) and nurture her (Gen 2:5). In verse 7, however, this is situation is rectified when God takes soil from the *Adamah* and moulds it into a figure called *adam*, a human being. It is important to note the strong linguistic connection between the *Adamah* and the *adam*.[22] Newsom (2000) writes that

22 Trible (1978) captures this play on words in her translation of 'earthling from the earth' or 'groundling from the ground' so as to preserve the link between *adamah* and *adam* (pp. 76–78).

'there is no more telling indication of the intrinsic relatedness of things than in the similarity of words'. To call the creature *adam* is to recognise its solidarity with Earth (p. 63). *Adam* is an Earth being. Humans were created from Earth for Earth.

In Gen 2:15, God gives the human the task of completing what was absent at the beginning of the story: the human has the task of greening *Adamah*:

> *Then the Lord God took the human and put him in the garden of Eden to serve (*abad*) and preserve it.*
>
> (Gen 2:15)

Here, the specific role of the human (the *adam*) is to 'serve and preserve' *Adamah*.[23] The Hebrew term *abad* is often translated as 'till' (NRSV), but it literally means 'to serve'. In the Bible, serving is what people did when they devoted themselves to a person or task – members of the court served the king (II Kings 25:24; Jer 25:11); priests and worshippers served their God in the temple (Exod 3: 12; Deut 11:13). In Gen 2:15, humans are to devote themselves to Earth by attending to Earth's needs (Habel, 2009, p. 69).

As the narrative progresses, Earth helps to create the animals of the fields and the birds of the air, thus establishing a kinship between these creatures and the human being. Earth becomes the common ancestor of all through the soil that is the basis for both human and animal life (Gen 2:7, 19). Brett (2000) detects a connection here between these verses and the genealogical formula in Gen 2:4a, which claims, 'these are the

23 Habel (2011) translates this phrase as 'to care for and conserve' (p. 49).

generations of the heavens and the earth' (p. 82). Here, Earth is the parent and all who dwell on Earth share a common origin.[24]

Within Genesis 2, then, humanity is depicted as being part of a single, unfolding reality. Human beings are not separated from Earth, but rather there is a deep continuity of life. This understanding resonates with contemporary scientific understandings of the human place within the cosmos. As Hillel (2006) writes,

> *This view of humanity's role accords with the modern ecological principle that the life of every species is rooted not in separateness from other forms of life in nature, but in integration with the entire living community.*
>
> (p. 243)

It is important to recognise that the image of relationship drawn here from Genesis 2 does not pretend to provide solutions to the intricate and complex issues specific to the climate crisis. Rather, this biblical vision (and others found elsewhere in the biblical text) provide faith communities with a foundation for re-conceiving the relationship between God, Earth and humanity, from which these communities might address the multiplicity of issues related to the ecological crisis. In emphasising

24 Some interpreters note that the naming of the animals by the *adam* denotes hierarchy and human dominance. In the biblical tradition, the process of naming serves a range of different functions. In Gen 16:13, for example, a new name seeks to articulate a new experience. In Ruth 4:13–17, the process of naming is a celebration of life and a reflection of communal connection. It is important to note that in Gen 2:19 there is nothing in the text that points to a hierarchical relationship between the *adam* and the animals.

mutual relationality, service and the subjectivity of all, this text provides an alternative to the anthropocentric outlook, which is often implicit in our liturgies, our worship and our ideas about justice and morality. Set alongside the ever-increasing scientific discoveries that point towards interdependence and the continuity of all life, this text (and others like it) has the potential to act as a powerful resource for Christian communities in the ongoing endeavour to alleviate environmental degradation.

REFERENCES

Balabanski, V., & Habel, N. C. (Eds.). (2002). *The Earth story in the new testament*. Sheffield: Sheffield Academic Press.

Berry, T. (1996). Into the future. In R. S. Gottlieb (Ed.), *This sacred Earth: Religion, nature, environment* (pp. 410–414). New York, NY: Routledge.

Brett, M. G. (2000). Earthing the human in Genesis 1-3. In N. Habel & S. Wurst (Eds.), *The Earth story in Genesis* (pp. 73–86). Sheffield: Sheffield Academic Press.

Burdon-Manley, L. (2019). India's Kerala struggles to recover from devastating floods. *Al Jazeera*, August 20. Retrieved from https://www.aljazeera.com/news/2019/08/indias-kerala-struggles-recover-devastating-floods-190819145122791.html

Code, L. (2006). *Ecological thinking: The politics of epistemic location*. New York, NY: Oxford University Press.

Collins, P. (1995). *God's Earth: Religion as if it really mattered*. Melbourne, VIC: Dove.

Darragh, N. (2000). Adjusting to the newcomer: Theology and ecotheology. *Pacifica*, *13*(2), 160–180.

Eaton, H. (2000). Guiding ecojustice principles: The Earth Bible Team. In N. Habel (Ed.), *Readings from the perspective of Earth* (pp. 38–53). Sheffield: Sheffield Academic Press.

Foster, A. (2019). Multiples fires merge together to create fire the size of Sydney. *News.com.au*, December 6. Retrieved from https://www.news.com.au/technology/environment/climate-change/multiple-fires-merge-together-to-create-fire-the-size-of-sydney/live-coverage/9148aee4001cd581fa5d5596759c8cca

Habel, N. C. (2000a). Introducing the Earth Bible. In N. C. Habel (Ed.), *Readings from the perspective of Earth* (pp. 25–37). Sheffield: Sheffield Academic Press.

Habel, N. C. (Ed.). (2000b). *Readings from the perspective of Earth*. Sheffield: Sheffield Academic Press.

Habel, N. C. (2001a). Is the Wild Ox willing to serve you? In N. C. Habel & S. Wurst (Eds.), *The Earth story in wisdom traditions* (pp. 179–189). Sheffield: Sheffield Academic Press.

Habel, N. C. (2001b). *The Earth story in the Psalms and the Prophets*. Sheffield: Sheffield Academic Press.

Habel, N. C. (2006). Playing god or playing Earth? An ecological reading of Genesis 1:26-28. In F. Gaiser & M. Throntveit (Eds.), *"And god saw that it was good." Essays on creation and God in honor of Terence Fretheim* (pp. 33–41). St Paul, MN: Luther Seminary.

Habel, N. C. (2008). Introducing ecological hermeneutics. In N. C. Habel & P. Trudinger (Eds.), *Exploring ecological hermeneutics* (pp. 1–8). Atlanta, GA: Society of Biblical Literature.

Habel, N. C. (2009). *An inconvenient text.* Adelaide, SA: ATF Press.

Habel, N. C. (2011). *The birth, the curse and the greening of Earth: An ecological reading of Genesis* (pp. 1–11). Sheffield: Sheffield Phoenix Press.

Habel, N. C., & Trudinger, P. (Eds.). (2008). *Exploring ecological hermeneutics.* Atlanta, GA: Society of Biblical Literature.

Habel, N. C., & Wurst, S. (Eds.). (2000). *The Earth story in Genesis.* Sheffield: Sheffield Academic Press.

Habel, N. C., & Wurst, S. (Eds.), (2001). *The Earth story in wisdom traditions* Sheffield: Sheffield Academic Press.

Hillel, D. (2006). *The natural history of the Bible: An environmental exploration of the Hebrew scriptures.* New York, NY: Columbia University Press.

Horrell, D. G., Hunt, C., & Southgate, C. (2008). Appeals to the Bible in ecotheology and environmental ethics: A typology of hermeneutical stances. *Studies in Christian Ethics*, *12*(2), 219–238.

Hurricane Dorian: Path of destruction. (2019). *BBC News*, September 9. Retrieved from https://www.bbc.com/news/world-latin-america-49553770

Masters, J. (2019). Hurricane Dorian was worthy of a Category 6 rating. *Scientific American*, October 3. Retrieved from https://blogs.scientificamerican.com/eye-of-the-storm/hurricane-dorian-was-worthy-of-a-category-6-rating/

McFall-Johnsen, M. (2019). Greenland's ice is melting at the rate scientists thought would be our worst-case scenario in 2070. *Business Insider*, August 15. Retrieved from https://www.businessinsider.com.au/greenland-ice-melting-is-2070-worst-case-2019-8

Mori, S. (2019). Why did typhoon Hagibis Shatter Japan's rainfall records? *NHK World – Japan*, October 16. Retrieved from https://www3.nhk.or.jp/nhkworld/en/news/backstories/714/

Newsom, C. A. (2000). Common ground: An ecological reading of Genesis 2-3. In N. C. Habel & S. Wurst (Eds.), *The Earth story in Genesis* (pp. 60–72). Sheffield: Sheffield Academic Press.

Oelschlaeger, M. (1994). *Caring for creation: An ecumenical approach to the environmental crisis*. New Haven, CT: Yale University Press.

Ruether, R. R. (1992). *Gaia and god: An ecofeminist theology of Earth healing*. San Francisco, CA: HarperSanFrancisco.

Scharper, S. B. (1999). The ecological crisis. In G. Baum (Ed.), *The twentieth century: A theological overview* (pp. 219–227). New York, NY: Orbis.

Schüssler Fiorenza, E. (1999). *Rhetoric and ethic: The politics of biblical studies*. Minneapolis, MN: Fortress Press.

Trible, P. (1978). *God and the rhetoric of sexuality*. Philadelphia, PA: Fortress Press.

Watts, J. (2019). Record heatwave 'make much more likely' by human impact on climate. *The Guardian*, August 2. Retrieved from https://www.theguardian.com/environment/2019/aug/02/record-heatwave-made-much-more-likely-by-human-impact-on-climate

White, L., Jr. (1967). The historical roots of our ecological crisis. *Science*, *155*, 1203–1207.

9

REACHING 'NET ZERO' – AN ENERGY PERSPECTIVE

Michael Colechin

With a shout and a chant the 'Moana – Water of Life' Conference opened, as the delegation from Oceania entered the conference theatre celebrating an activity that underpins their culture – working together to paddle seagoing canoes. These island communities live in harmony with the oceans; oceans that connect them, feed them, provide them with cultural identity and now threaten to engulf them as the result of the actions of people living thousands of miles away. Climate change is a very real, everyday fact of life for people in these communities living with the knowledge that, as sea levels rise and extreme weather events become more frequent, it is very likely that their homes will become uninhabitable within their own lifetimes. The stories each of them told created a compelling personal narrative around the need to ameliorate global warming. They also raised the very real consequences of the damage that climate change is doing – creating mass migration and communities of refugees larger than any seen since World War II, with no hope of ever returning to their homes.

A few years ago I spent a holiday with my family building a wooden canoe that we now paddle on a regular basis on the rivers near our home. A wooden canoe built for paddling Britain's relatively benign waters is as nothing compared to seagoing craft designed to tackle Pacific Ocean waves. I only have the vaguest concept of the connection the people of Oceania feel to one another and to the seas on which they travel. And yet, I believe that if we are to address the challenge of climate change in ways that deliver solutions such as becoming 'Net Zero' in our greenhouse gas emissions, this is the level of ambition and connection to one another we will need.

Approaching climate change as an engineer, a problem-solver, it is hard not to be drawn into a technological framing and a belief that the problem can be overcome with appropriate use of low carbon and renewable energy supplies. If only we built enough wind turbines and solar panels and deployed other low-carbon technologies, surely we could create an abundance of clean energy. However, using electricity is only one of the ways in which we consume energy. Heating our homes and businesses, and transporting ourselves and the goods that we purchase, all use far more energy than is currently produced by our electricity system. It will be possible to 'decarbonise' the electricity system by 2050 (the date currently set as the UK's objective for achieving Net Zero), and with a heroic effort (well beyond that currently being made) we may even be able to do it well before this date. However, we will not in those same timescales be able to expand the electricity system to the level whereby it is not only delivering our current demand for electricity but also our demand for heat and transport. Even if we can build the vast quantities of additional renewable electricity generation capacity that would be needed, an all-electric energy system would require us to electrify the whole of our transport system, and the heating system in every individual business and dwelling in the country would have to be changed so that they could run on electricity.

Other low-carbon solutions for heating and transport are being explored, and if we are to be successful at achieving our decarbonisation goals we will need to use all of these solutions. In this context the challenge becomes one of optimising the portfolio of solutions available – a complex set of interactions where the goals are not only environmental but also commercial and societal. The most effective technical solutions are not necessarily those that can be afforded, nor will they always deliver jobs or social equity.

This calls for us to frame the question in a different way. Rather than just asking ourselves how we can decarbonise our energy supplies, we need to be considering how we can change the whole system to make our use of energy more sustainable. Fundamentally, this requires a reduction in our demand for energy – demand that is driven by the decisions we all make on a daily basis.

When looking at this challenge there is a temptation to consider the problem as one of efficiency – if you can reduce the amount of energy needed by homes and businesses by making them more efficient, energy demand will be reduced. There is certainly merit in this response, and much good work continues to go on to improve the insulation of our homes and reduce industrial energy consumption. However, our economic structure – the drive for continuous growth – encourages us to use the capacity released by efficiency improvements to support further growth. Whilst it can be argued that the efforts made in recent years have broken the link between economic growth, energy consumption and emissions, there is a long way to go before we reverse the trend, creating a functioning economy that is Net Zero in its emissions of greenhouse gases.

Economists might argue that just such a situation can be achieved through a fully functioning market. After all, energy can be considered to be a commodity that can be bought and sold, saved or wasted. Controlling energy demand and the emissions it creates may simply be a case of putting in place the right economic incentives and penalties. Unfortunately, the increasing complexity of the energy system mitigates against this, incentives in one area become a penalty in another and there are many opportunities for the use of such measures to create unintended consequences.

Ultimately, it is not energy that matters but the services it provides, such as comfort, mobility, lighting or the means of production. The need for these changes across cultures and time as societies change (The DEMAND Centre, 2018). Steering and shaping these transformations is arguably the best chance we have for reducing the greenhouse gas emissions that lead to climate change.

Food production is one of these services. Reducing food waste has long been recognised as a route to reducing greenhouse gas emissions, and with around 25% of the food produced in the UK being wasted there is a lot to go at. Each of the 10 million tons of food that we waste each year in the UK leaves behind a legacy of avoidable greenhouse gas emissions. These emissions are not only associated with the production of the food itself but also the energy required for its subsequent transportation, storage (often requiring refrigeration) and sale. In the UK the emissions associated with food wastage have been estimated to be equivalent to around 5% of the country's total greenhouse gas emissions. However, when you also consider the food we don't waste there is huge potential to further reduce our emissions through the decisions we make on a daily basis. Choosing to eat less meat, purchasing locally sourced and seasonal food, converting to a vegetarian or even vegan diet are all options that have been shown to reduce greenhouse gas emissions.

We also make daily decisions about the way we get around, and these too have a huge impact on greenhouse gas emissions. If we choose to travel on our own in a vehicle powered by fossil fuels we will create more CO_2 emissions than if we were to share that journey with others, use public transport, cycle or even walk. It is not just our own decisions that matter here; we can all play a role in encouraging others to make different choices. In the organisations in which we work and/or volunteer, this may include requesting cycle changing facilities, discussing the health benefits of walking and/or supporting those who decide to travel by train instead of driving. This is particularly important where we have leadership roles. We can change cultures such that choosing the lowest carbon option (reducing the cost to the planet) is considered as important as controlling financial cost.

Choosing to reduce our emissions needn't always have to cost more. Lower carbon options can reduce financial as well as environmental costs. In business, with the aid of new technologies, we don't always have to travel to meet colleagues, and where we have responsibility for investment decisions we can choose the more energy efficient options that save money as well as reducing our emissions. Similarly, we can save money at home by changing some of our household practices. Turning lights off, reducing the temperature setting on our heating systems, and air-drying rather than tumble-drying our clothes all reduce our greenhouse gas emissions and save us money.

It's not just how we wash our clothes that's important but also the choices we make when buying them. Today the fashion industry is dominated by 'fast fashion'. This business model has also ratcheted up the industry's environmental impact. The fashion industry is now one of the top five polluters in the world. Thinking about the environmental impact of our clothing might lead us to choose to repair rather than replace, and only buy what we need from producers who are committed to making a difference. When we do buy new clothing, the emissions created as the goods are transported are also important. Some of the practices that have built up around online shopping, like having multiple items delivered and sending back those that don't fit, have added significantly to the environmental impact of the industry.

Finally, our big life decisions also have a significant impact on our personal demand for energy and our impact on the environment. The choices we make about where to live and work have the potential to significantly alter our environmental impact. Taking climate change into account as we make these decisions may lead us to buy smaller or more efficient homes or to choose to work for very different organisations.

The lesson in all of this is that if we take people's behaviours for granted, if we design the future based on what has happened in the past, the solutions we adopt may not be effective. At worst they may even increase our demand for energy and the greenhouse gas emissions associated with its production. The whole of society needs to be engaged in thinking that energy is precious. For too long we have considered it an abundant resource. Energy is all around us, but the methods we currently use to harness it have a penalty which is currently being borne by others, often many thousands of miles away in communities very different from our own. This penalty will ultimately be borne by all nations, societies and communities.

REFERENCE

The DEMAND Centre. (2018). What is energy demand? And why does it matter. The DEMAND Centre official website. Retrieved from http://www.demand.ac.uk/. Accessed on March 3, 2020.

10

SUSTAINISLANDHOME.ORG: AN APP FOR CLIMATE ACTION AND ADVOCACY

Sheila Moore Andrus

INTRODUCTION

"Sustain Island Home" is a phrase from the Episcopal Church Book of Common Prayer wherein God is praised for the blessings of "...this fragile earth, our island home....".[1] Though published more than 40 years ago, the authors of the prayer now seem prescient in lifting up the fragility of the earth, the idea of earth as "home" and the reality of, for now, our only home. We who were gathered in Lincoln for the *Moana Water of Life Conference* – scientists, theologians, practitioners, students – each brought to the conference our awareness of the blessings of the earth, as well as the

1 *The Book of Common Prayer, Holy Eucharist II, Eucharistic Prayer C* (p. 370). (1977). New York, NY: Seabury Press.

unfolding of the long emergency of the climate crisis that affects all life now, and will continue to do so for many generations to come.

The climate crisis is frightening for many people because of evident and often dire impacts, such as rapid species loss, ecosystem breakdown across lands and seas, rising ocean waters, droughts, fires, floods, food insecurity, spread of vector-borne diseases, political strife, and economic disruptions, among others. Although these impacts are alarming, it's often not clear what one person, or one community, can do to drive down greenhouse gas (GHG) emissions causing the problems. For this reason, I was honored to present information at the Moana Conference about a hopeful effort that I'm helping lead for the Episcopal Church; that is, the rollout of an Internet accessible tool – sustainislandhome.org – that can help people and communities be part of climate solutions.

My presentation, summarized in this chapter, focused on why sustainislandhome.org was developed, its design principles and how it works, and lessons we are learning from our work with the Episcopal Church. It's my hope that our effort can serve as a model for other faith communities in educating and mobilizing their members for climate action and advocacy.

WHY DEVELOP A WEB-BASED PLATFORM, SUSTAINISLANDHOME.ORG, FOR THE EPISCOPAL CHURCH?

The climate crisis due to global warming demands a worldwide response, especially from people of faith who believe that the earth is a sacred gift from God. Impacts from global warming are happening now and more quickly than anticipated. The 2018 Intergovernmental Panel on Climate Change (IPCC) report predicts that many impacts previously anticipated to occur at 2.0 ° C global warming above preindustrial levels are now projected at 1.5 °C by 2040. To mitigate catastrophic impacts, the IPCC scientists recommend a cut in global GHG emissions by 45% by 2030, and a reduction to zero by 2050 – goals that can only be reached with a mobilized global community understanding that "... every half degree matters, and every action matters" in reversing GHG emissions driving global warming.[2]

Social justice also compels people of faith to climate action because too often the people that are the first and hardest hit are also the least responsible for the climate crisis. As an example, an estimated 20–40 million people per year are currently environmental refugees, fleeing rising waters, droughts, storms, and conflicts – a number estimated to increase to approximately 200 million people within the next 30 years. We understand, too, that for every 1 °C in global average temperatures, there is a 10% reduction in the world's grain crops; and, the World Bank estimates climate-related disasters are pushing 26 million people into poverty each year.

I described how broad measures of global warming impacts can be personal; for example, loss and damage in my home state of California in 2018 included two wildfires resulting in more than 90 deaths, 153,000 acres burned, and 14,000 homes destroyed. Resulting air quality in San Francisco, the city where I live, ranged from unhealthy to hazardous over a 2-week period, requiring the use of air filtration masks to go safely outdoors. Other, ongoing impacts in California include the warming and acidification of ocean waters, the erosion of coastlines, and the desperate movement of migrants from Central America across borders to escape food insecurity from droughts.

---

2 The Summary for Policymakers of the Special Report on Global Warming of 1.5 °C (SR15) is available at https://www.ipcc.ch/sr15 or www.ipcc.ch.

It was in this context of ever-increasing and rapid global warming impacts that I welcomed the opportunity to be part of a project sponsored initially by the Episcopal Diocese of California, under the leadership of the Rt. Rev. Marc Andrus, and later endorsed by the wider Episcopal Church in Resolution C008 at the 79th General Convention.[3] Beginning in January 2018, I worked with Climate Solutions Net and a team of staff and volunteers in the Episcopal Diocese of California to customize a web-based platform, sustainislandhome.org, that functions as an "app" designed to[4,5]:

- educate Episcopalians about actions they could take to reduce their household GHG emissions;
- connect Episcopalians to resources in their congregation, diocese, and across the Episcopal Church for learning, partnering, and actions in keeping with a more sustainable life; and,
- amplify the voice of Episcopalians in policy arenas by showing the amount of GHG emissions reduced at the level of church, diocese, and Episcopal Church. This last feature gives both a visible and encouraging witness of environmental and justice commitments, as well as demonstrating the power of community to effect change.

---

3 C008 Advocacy for Creation Care (https://www.vbinder.net/resolutions/C008?house=hd&lang=en).

4 Climate Solutions Net (https://www.climatesolutionsnet.com/), a California-based company that provides a "…SaaS platform…to drive climate solutions engagement…."

5 Definition for "app": "Short for application, the commonly used abbreviated form ("app") refers to a specific type of application that is designed for a single purpose. An app has a single functionality whereas an application may handle a number of functions." https://www.webopedia.com/TERM/A/app.html.

Underpinning our efforts was not just the urgency for each person to be part of climate solutions, especially in the next 10–15 years, but also the fact that options are available now to households for a more sustainable lifestyle that conserves energy and water. A key encouraging statistic is that five basic household activities make up 40% of GHG emissions in the United States: home electricity use, local transportation choices, air transportation, food choices, and recycling. This means that, beyond any actions by industry or agriculture, choices by households can make a significant impact on climate change. The challenge then becomes educating people about the range of options available to them so they can begin to make changes for a more sustainable life, and sustainislandhome.org addresses this need.

Using sustainislandhome.org, Episcopalians can learn about options and develop a plan for their home that makes a real impact on GHG emissions, and the cumulative choices of Episcopalians add up: for example, 3 million Episcopalians, or 1.5 million households can reduce on average 4–5 tons CO_2e/year, or 6–8 MT (MT = megatons or million metric tons). These reductions are comparable to emissions produced by Coca Cola in manufacturing each year, or emissions produced by many less developed countries, such as Uganda (5.2 MT CO_2e), El Salvador (6.3 MT CO_2e), Cambodia (6.7 MT CO_2e), or Costa Rica (7.8 MT CO_2e), among others.

In addition to helping Episcopalians drive down their GHG emissions, this project is also relevant to advocacy for policy at many levels, including at the level of the United Nations. As acknowledged by the former Archbishop of Canterbury, Rowan Williams, it takes a global body to deal with a global problem. The Episcopal province in the Anglican Communion is part of the worldwide Anglican Communion. Comprised of 17 member states of the United Nations, the Episcopal Church has an official observer status at the United Nations Framework on Climate Change Conference of Parties.[6] This means that the Episcopal Church is able to advocate for Episcopal Church policies to UN negotiators, such as the cap for 1.5 °C warming, promoting renewable energy technologies, supporting a just economic transition for the most vulnerable world populations, and promoting tools like sustainislandhome.org that can aid in climate action.

In summary, the "why" of the sustainislandhome.org was for all the reasons above – people of faith understand the need to join the global community in climate action, the fact that sustainable energy options are available to support household action, and the need and opportunity for people of faith to support not only individual behavioral change but also community advocacy for care of the Earth and the most vulnerable people and species.

HOW DOES SUSTAINISLANDHOME.ORG WORK, AND WHAT ARE THE KEY DESIGN PRINCIPLES?

To best understand how sustainislandhome.org works, I recommend watching a video demonstration available at diocal.org/climate. The video gives an overview of sustainislandhome.org; that is, how to register for use and its online menu for viewing actions and one's progress in decreasing GHG emissions, conserving water, and cost savings for utilities. A brief summary of how sustainislandhome.org works follows.

Once a user finds sustainislandhome.org, on his/her smartphone, tablet, or computer, they'll be taken to a web page that immediately gives messages from Episcopal Church

6 Dioceses of the Episcopal Church, described in anglicancommunion.org.

leadership about care of creation, metrics about wider church and diocesan-level actions that address climate change, and links to resources that are educational and inspirational. This page also has a prominent "Get Started" button at the top and bottom of the web page. Once the user clicks on this button, there are only three steps to becoming a climate champion: (1) registering to use the "app" by signing in with one's email and creating a password, (2) answering questions that establish one's energy baseline, and (3) using the software to learn about ways to reduce energy and water usage and then selecting actions for your household.

There are many helpful features to support these three steps; for example, the software leads the user through a series of questions that establishes their energy profile, and, with this baseline in place, the user can view a personal dashboard generated for their household that displays their GHG emissions per year. The dashboard also shows the household's changing CO_2e/year in real time, as actions are taken to drive carbon down, and displays summary information that includes comparison of one's GHG emissions to other homes in the same city, in the United States, and in a number of select developing countries.

In designing sustainislandhome.org, we sought to increase hope and inspiration, as well as educate the user to make changes in their lives, and harness the power of community for mutual support and advocacy. We emphasize a personal approach in introducing people to the tracker, both in online trainings via webinars announced on diocal.org/climate, and in presentations to groups. Successful engagement with this "app" depends on personal connections, beginning with providing an opportunity for people to talk about why they are concerned about the climate crisis and how climate action fits with shared values. Once the "app" is introduced to a church or group, engagement is strengthened when the group comes back together several times to continue their conversation, sharing tips about actions taken in their households and celebrating the community's success taking actions that that address the climate crisis.

LESSONS LEARNED

In rolling out sustainislandhome.org across the Episcopal Church, we've learned that climate grief and despair is a barrier to action among people of faith. Accordingly, our approach for introducing this "app" is as important as how this web-based tool works. By designing a tool that is available on the Internet, the educational process can be quickly available to many users; and, by introducing the tool in a personal way, a bridge is established for the future. The circles of conversation about an online tool become the circles of hope, inspiration, and action that can continue.

With 5–7 billion of Earth's population identifying as religious today, what the Episcopal Church is doing to model individual change, supported by community, can have far-reaching effects as other faith bodies consider ways to give hope to and educate their members on the need to address climate change. In this example from the Episcopal Church, we've used a web-based tool introduced in personal ways as an approach to mobilize climate solutions and advocacy of policies. For those leading climate action in the Episcopal Church, it is our prayer that our efforts will not only help reduce inputs causing global warming but also encourage other people of faiths to consider how they might mobilize the hope, education and actions of their members, so that together, we can sustain this precious Earth, our island home.

11

FAITH IN ACTION AT THE UNITED NATIONS: A CONFLUENCE OF RIVERS

Lynnaia Main

INTRODUCTION

Climate change is the defining urgency of our day, and increasingly governments are declaring a climate emergency. This is also true at the United Nations (UN). For most of us, the UN probably seems an ocean away, a place where world affairs are decided that sounds quite complex in how it relates to the regular person or local organization. Yet, it relates to us very directly and, no matter where you are on the planet, you can participate. Those of us who work to address climate change from the viewpoint of our faith are playing important roles there, and there is important future work to be done. This chapter explores the question: How can we, as faith-based participants, engage with the UN on climate action in smart, strategic ways in an era of climate emergency?

To answer this question, we will first examine the place of faith and the ways in which faith-based actors and organizations engage with the UN, using the metaphor of a confluence of rivers. We will look at their engagement in general focusing on climate change; the successes, challenges, and obstacles they have encountered; as well as the lessons learned for future faith engagement at the UN. We will then turn to our strategic reflection: based on these lessons learned, in a time of climate emergency, how can faith-based actors and organizations engage – or engage more deeply, strategically, and intelligently – with the UN system to address this emergency? The research and ideas presented here are preliminary – a work in progress – as they call for your response.

THE UNITED NATIONS: A CONFLUENCE OF RIVERS

Allow me to start by proposing to you the metaphor of faith in action at the UN as a confluence of rivers and inviting you to consider as illustration the confluence of rivers Baker and Neff in Chile – of which many photos and videos are available on the Internet – in which two distinct bodies of water join, mingle, mix, swirl, and become a new body in the process. At the UN, we might visualize three distinct rivers – member states, UN entities, and civil society – that meet, mingle, and collaborate on issues, influences, beliefs, practices, cultures, and actions.

Let us review some UN basics to understand this confluence. The UN was founded in 1945 when 50 nations signed the UN Charter in order to maintain international peace and security through global dialogue and action. Today, its role has expanded to five pillars: maintain international peace and security, protect human rights, deliver humanitarian aid, promote sustainable development, and uphold international law. According to the Overview (2019) on the UN's main webpage, as an international organization, its members and constituents currently comprise some 193 member states, whose actors include government officials, policymakers, and representatives. The UN was and is composed of six main organs and numerous UN funds, programs, and specialized agencies, collectively referred to as UN entities, whose actors include UN staff and diplomatic officials. The UN Charter's Article 71 foresaw a third group, collectively known as "civil society," consisting of "nongovernmental organizations" (NGOs) who benefit from two formal means of association. NGOs can be subdivided into various currents, such as nonprofit associations, for-profit organizations, academia, and faith-based organizations.

THE RIVER OF CIVIL SOCIETY: A BRIEF HISTORY OF FAITH IN ACTION AT THE UNITED NATIONS

Having reviewed the UN's raison d'être and structure, let us examine the role of faith through a brief history of faith in action at the UN. Often hidden and implicit, faith has been present, part of, and diffused within the UN system from its beginnings, informally and unofficially, through "faith-inspired" or "faith-based" individuals whose actions are rooted in their faith values. Early examples include the UN Conference of San Francisco delegates who, in 1945, worshipped at Grace Cathedral during the conference; Eleanor Roosevelt, an Episcopalian who served on the first US delegation to the UN, chaired the UN Commission on Human Rights and drafted the Universal Declaration of Human Rights; and Dag Hammarskjöld, the second UN Secretary General whose book *Markings* reveals meditations on faith and who designed the Meditation Room at UN headquarters to be its spiritual center. Faith values also find their expression in formal, organized, institutionalized ways through the relationships the UN fosters with "faith-based organizations," also known as "religious nongovernmental organizations." These might be singular religious institutions or denominations (such as the Holy See, Greek Orthodox Church, Anglican Communion, The Episcopal Church), multifaith and interfaith organizations (such as the World Council of Churches, Religions for Peace or Parliament of the World's Religions), or faith-based working groups and coalitions (such as the Committee on Religious NGOs, NGO Committee on Spirituality, Values, and Global Concerns, Ecumenical Women, or the NGO Committee on Freedom of Religion and Belief).

To return to our confluence metaphor, if civil society is a river, faith is one of its currents, running sometimes stronger, sometimes weaker, even as the river itself ebbs and flows. Despite early faith foundations, between 1945 and 2000 the UN traversed what might be considered a period of allergy characterized by disinterest – sometimes aversion – to working with faith-based actors, who were neither intergovernmental, nor secular, nor interested, usually, in working with the UN. In the 1990s, allergy gradually warmed to embrace, as more NGOs engaged with the UN system and UN agencies – UNICEF, UN Population Fund (UNFPA), UN Development Programme (UNDP), UN Environment Programme (UNEP), to name a few – discovered the utility of working with faith-based actors as local partners with deep community roots and communal resources and meeting spaces. The year 2000 marked a turning point, when 1,000 religious leaders gathered at the UN, invited by then UN Secretary General Kofi Annan who held great interest in the role of faith-based actors in the world, for the Millennium World Peace Summit of Religious and Spiritual Leaders, to deliver a declaration demonstrating their support for eradicating poverty by 2015 (*Millennium World Peace Summit of Religious, Spiritual Leaders Meets at Headquarters*, 2019). This largest ever gathering of religious leaders at the UN was meant to identify ways that the worldwide religious and spiritual communities could work together as interfaith allies with the UN on specific peace, poverty, and environmental initiatives (Kofi Annan Faith Briefing Agenda, 2019c). Their noteworthy presence and support expanded this new embrace and appreciation of faith-based actors. The next two decades saw the creation of the UN Inter-Agency Task Force on Religion and Development (IATF-FBOs) in 2010 and which, by 2018, incorporated 19 UN agencies working with faith-based organizations (UN Interagency Task Force on Religion and Development – Annual Report, 2018), and the Multi-Faith Advisory Council in 2018, consisting of religious leaders who consult with UN agencies on common goals.

Ebb and flow, allergy and embrace describe well the movements and sentiments of and toward faith-based organizations in their UN engagement. This dynamic, fluid tension affects their interest, propensity, and ability to engage with member states, UN agencies, and each other in the confluence that is the UN.

FAITH-BASED ENGAGEMENT ON CLIMATE CHANGE AND WATER AT THE UN: SUCCESSES, CHALLENGES, AND OBSTACLES

While faith-based engagement with the UN has been constant, engagement on climate change and water has been more recent, as the UN itself has continued to develop in response to climate change. Let us consider the means of engagement and the successes, challenges, and obstacles they have encountered.

The UN's architecture for addressing climate change dates to the 1990s with a sequence of international legal frameworks and mechanisms including the UN Framework Convention on Climate Change (UNFCCC) (United Nations, 1992), the Kyoto Protocol (United Nations, 1997), and the Paris Agreement

(2015). Other frameworks relate indirectly, such as the Conventions on Biological Diversity (United Nations, 1993) and to Combat Diversification (United Nations, 1994), Agenda 2030 (2015), and the Global Compact on Migration (United Nations, 2018) (*Climate Change*, United Nations, 2019). No single UN agency serves as climate change focal point, but an increasing number are involved: the UNFCCC Secretariat, the Intergovernmental Panel on Climate Change (IPCC), UN Environment, UN Water, UNICEF, and UNDP, all of whom focus on reducing greenhouse gases and carbon emissions, pollution, and loss of biodiversity. Two recently formed UN Environment initiatives collaborate closely with faith actors, the Faith for Earth Initiative and the Interfaith Rainforest Initiative.

As for water, the traditional UN focus on WASH – Water, Sanitation and Hygiene – has prioritized access to clean, safe, protected water; water as a human right; humanitarian assistance; and eradicating poverty rather than serving as a locus of climate action, in part because of its segmentation in a separate section of the UN system. However, anecdotal evidence suggests that some UN agencies, traditionally unconcerned by climate change, are focusing on its impact on water. For example, at UN side events on climate change and water in July 2019, UNICEF and UNESCO speakers talked about rethinking their water focus to account for and incorporate climate change.

Within these frameworks, faith-based actors and organizations engage with member states, UN entities, and each other in several ways. They raise awareness of faith-based values on caring for creation and the planet by publishing references to sacred scriptures, such as the papal encyclical *Laudato Si': On Care for Our Common Home*; by issuing multifaith declarations; by organizing and participating in pilgrimages, protests, marches, and mass mobilization; by fostering interfaith conversation, chant, and prayer; and through storytelling.

They also participate formally in UN meetings and negotiations. Before UN meetings, preparatory actions might include networking, information sharing, bilateral and multilateral advocacy. This is followed by formal participation during UN meetings as observers, participating in advocacy bilaterally or in coalitions, and addressing meetings through oral or written interventions. They may lobby during the negotiations, organize and participate in parallel events, and hold prayer and worship services. An example of this action was the organization of daily interfaith prayer in the public area of the UN space by Episcopal Church informal delegations at several United Nations Climate Change Conferences, also known as the Conferences of Parties (COPs), between 2015 and 2018.

Finally, a more marathon-like means by which faith-based actors engage with the UN is by nurturing relationships and partnerships with UN actors. This means networking with member states and their missions – such as the mission visits and communications with the governments of Samoa, Fiji, and the United Kingdom conducted in preparation for this presentation – UN staff and civil society organizations; elevating faith-based voices, stories, and experiences; contributing technical expertise such as policy reports; and partnering with UN entities. This ministry of presence, whether physical or virtual, lets the world know they are there, and they care.

While potentially boundless, the issues and themes raised by faith-based actors and organizations commonly refer to foundational values and texts expounding on the sacredness of life and the multifaith values of caring for creation; ensuring justice for those most vulnerable and at risk; human rights to water and land; the faith-rooted mandate to welcome the stranger; and the call to simplicity and reduction of consumption, greed, and selfishness.

Having explored this simple typology of the ways faith-based actors engage with the UN on climate change and water, let us consider the results. How successful have they been, what obstacles and challenges have they met? To address these questions, I conducted documentary and qualitative research during the spring and summer of 2019. My research methodology included (1) documentary reports and research such as academic articles, news articles, and UN reports; (2) personal experiences and observations; and (3) interviews with several representatives of faith-based organizations, member states, and UN agencies. It must be stressed that the findings presented here are limited in scope and statistical representativity; nevertheless, the anecdotal evidence pointed to several successes, challenges, and obstacles to faith-based/UN engagement on climate change.

Among the successes suggested by the research is the fact that faith values are more widely known now within the UN system, as major declarations of faith – such as *Laudato Si'* – have been embraced by UN and secular organizations alike. Faith-based actors also are recognized as bringing ethical and moral leadership to environmental, ecological, and climate change concerns, demonstrated by their increasingly formalized representation via an Interfaith Liaison Committee that communicates with the UNFCCC Secretariat. Indigenous leaders are recognized for their spiritual wisdom and traditional technical expertise. Faith actors are recognized for holding governments and businesses to account, mobilizing communities, speaking up for those without voice and for "niche" areas of expertise, such as the role spiritual leaders can play in dealing with climate grief, and the stewardship, hope, and renewal that religious leaders convey through teaching and preaching. Finally, faith-based actors are respected for their real connection and influence on their constituencies and empowerment of their local communities. The Anglican Communion, for example, has tremendous influence with around 85 million members in 165 countries.

The research also identified several challenges and obstacles. Like all NGOs, faith-based actors have a limited scope of action and influence because of the constraints of the UN system. They do not engage entirely on their own terms, but rather according to UN criteria and have no named constituency or major group representing them at the UN Climate Change Conferences. It takes a great deal of time, money, and willpower to navigate the gigantic UN bureaucracy; civil society space is shrinking at the UN and rising populist nationalism has challenged multilateralism's legitimacy. Faith actors also have internal limitations: meager financial and staff resources, and lack of awareness, apathy, or distrust of the UN. They are at risk of being instrumentalized and must be aware of this as they engage with the UN. Further, there is the difficulty of measuring "success": How can we know whether our mission visit or other action is going to change a decision-maker's mind? There is an element of faith in doing this work; it is almost impossible to quantify influence, and often the influence might be within ourselves, transforming our understanding of our importance in the world as people of faith. Finally, faith-based actors must cope with the denial, doom, and distance factors that commonly inhibit climate action.

LESSONS LEARNED: STRATEGIC ENGAGEMENT IN AN ERA OF CLIMATE EMERGENCY

We have examined faith-based engagement at the UN on climate change and water; and some of the successes, challenges, and obstacles reported in the literature and via interviews. Climate change has, until recently, been an emerging issue moving from the margins to the center of UN dialogue and action. Today, climate change is not only a central issue, but increasingly the number one priority at the

UN, overshadowing all other priorities. The urgency was underlined in the Intergovernmental Panel on Climate Change's *Special Report on Global Warming of 1.5 C* (2018) that gave a 12-year timeline for averting the tipping point. More recent reports have reduced the timeline to only 18 months (McGrath, 2019). The urgency of the climate emergency was highlighted in a video, *Secretary-General António Guterres calls for global action on climate change* (2018, Sept 10), ahead of the SG's Climate Action Summit in 2019. The race to manage climate change's catastrophic effects is not a marathon; it has become a sprint.

Bearing this urgency in mind, what lessons might faith-based actors draw upon to develop smart, strategic engagement, as the luxury of time – since relationship building is a marathon – slips away? The research for this presentation suggested at least seven lessons:

First, stay at home, stay local, reduce your carbon footprint, and modify your individual and household consumption patterns, using mitigation tools such as the carbon tracker promoted by the Episcopal Diocese of California. Second, shift focus from advocacy and negotiations to local solutions through mitigation and adaptation. Third, push for local and national implementation of the Paris Agreement and all other relevant UN agreements. Fourth, massage the advocacy message so that it feels do-able, recognizing that people are not necessarily ready to engage at the same levels. Fifth, prioritize interfaith actions and coalitions (the preferred means by which the UN engages with faith actors), youth action, and mass mobilization. Sixth, choose a means of UN engagement that will be effective and appropriate for your organization's size and find UN connectors who can help you engage. Seventh, engage the UN from where you are, either locally and virtually or more broadly, by connecting with UN headquarters in New York and Geneva or at UN regional offices. UN representatives of faith-based organizations serve as connectors to this complex world and can customize ideas for curating your engagement with the UN.

The lessons listed here are incomplete, as they require your input and action. The tools are starting points, proposed by those who already are involved and have thoughtfully offered their input.

CONCLUSION

Whether we are a faith-motivated individual, a faith-based organization, or simply inspired to keep faith and hope alive for the future of humanity, climate change is our call to action. As people of faith, people of science, people of the planet, we are called to care for creation and our planet, to work and act in concert. Climate change affects us locally, and each of us has stories and solutions to share based on local resilience. To meet the necessary levels of ambition, we also need global solutions. And while the UN is not the only confluence of global dialogue and action, it is a unique and fruitful arena that deserves our attention.

Nothing stands in the way of us acting globally with the UN. We are living in a time of growing awareness and appreciation of the role of faith-based actors, and this presentation hopefully has demonstrated some of the myriad ways in which faith-based actors and organizations have been and continue to address climate change and water concerns in the UN context. While faith-based actors already have contributed much within the UN system, they – and we! – are well poised to contribute uniquely and fully. And this, no matter one's place on the planet.

Let us link our call to action with a message of hope. In September 2019, young Swedish climate activist Greta Thunberg arrived in New York City by sailboat, having traversed the Atlantic to attend the UN Secretary General's

Climate Action Summit. Photos posted to her Twitter account (Thunberg, 2019), taken aboard her sailboat as she arrived in New York Harbor (a confluence of rivers), show the New York City skyline behind her and the Statue of Liberty before her. The photo of the Statue of Liberty "welcoming" Greta and her crew is powerfully symbolic, as this monument – the beloved "Mother of Exiles" – is iconic for its welcome of refugees and immigrants. At its base, a plaque is inscribed with a poem proclaiming what many consider America's best spirit of welcome and inclusivity: "Give me your tired, your poor, Your huddled masses yearning to breathe free, The wretched refuse of your teeming shore. Send these, the homeless, tempest-tost to me" (Lazarus, 2019). We will need to depend increasingly on that spirit in future. Certainly, a central message relayed by participants at this conference has been the present and future importance of welcome, and how much it is part of all our faith traditions.

People of faith and people of hope: this is our call to action. We will need to welcome. We will need to teach others, by word and example, about radical welcome and hospitality. So that we do not build walls, so that we receive those in need. My prayer is that we will view the UN as a tool enabling us to connect globally, and our faith traditions as a primary teaching of what welcome looks like.

A very simple prayer to encourage us in that action comes from that pioneering man of faith at the UN, Dag Hammarskjöld (1964): "...For all that has been – Thanks! To all that shall be – Yes!" (p. 89).

REFERENCES

Anglican Alliance. (2019, February 14). People forced from home by climate change: Lambeth Roundtable on Climate

Change Migrants and Refugees – with a focus on small island states. Retrieved from https://anglicanalliance.org/people-forced-from-home-by-climate-change-lambeth-roundtable-on-climate-change-migrants-and-refugees-with-a-focus-on-small-island-states/.

Bassett, L., Brinkman, J., & Pedersen, K., (Eds.). (2000). *Earth and faith: A book of reflection for action.* New York, NY: United Nations Environment Programme Interfaith Partnership for the Environment.

Baumgart-Ochse, C., & Wolf, K. (2018). *Religious NGOs at the United Nations: Polarizers or mediators?* Abingdon, OX; New York, NY: Routledge.

Bird, F., Twiss, S., Pedersen, K., Miller, C., & Grelle, B. (2016). *The practices of global ethics: Historical backgrounds, current issues and future prospects.* Edinburgh: Edinburgh University Press.

Carrette, J., & Miall, H. (Eds.). (2017). *Religion, NGOs and the United Nations: Visible and invisible actors in power.* London: Bloomsbury Publishing.

Cranston, R. (1968). *World faith: The story of the religions of the United Nations.* Freeport, NY: Book for Libraries Press.

Disney Wiki. (2019). Moana. Retrieved from https://disney.fandom.com/wiki/Moana. Accessed on April 2, 2019.

EcoSikh (2016, August 22). Sikhs join faith leaders handing interfaith climate change statement to United Nations. *Sikh24.com.* Retrieved from https://www.sikh24.com/2016/04/22/sikhs-join-faith-leaders-handing-interfaith-climate-change-statement-to-united-nations/#.XVTd1UF7nyc

Glaab, K. (2017). A climate for justice? Faith-based advocacy on climate change at the United Nations. *Globalizations, 14*(7), 1110–1124. doi:10.1080/14747731.2017.1308060

Grace Cathedral. (2019). People and Initiatives. Retrieved from https://gracecathedral.org/people-initiatives/

Hammarskjöld, D. M. (2006, originally 1964). *Translated from the Swedish by Leif Sjöberg & W. H. Auden.* New York, NY: Vintage.

Haynes, J. (2014). *Faith-based organizations at the United nations*. New York, NY: Palgrave Macmillan.

Human Rights Commission. (2019). The Eleanor Roosevelt papers project. George Washington University. Retrieved from https://www2.gwu.edu/~erpapers/teachinger/glossary/human-rights-commission.cfm

Intergovernmental Panel on Climate Change. (2018). Special report on global warming of 1.5 C.

Johnston, D., & Sampson, C. (1995). *Religion, the missing dimension of statecraft*: Oxford University Press.

Kent, M. (2018, March 22). Canadian teen tells UN 'warrior up' to protect water. *CBC News*. Retrieved from https://www.cbc.ca/news/canada/autumn-peltier-un-water-activist-united-nations-1.4584871

Lazarus E. The new colossus. National Park Service. Retrieved from https://www.nps.gov/stli/learn/historyculture/colossus.htm

Lehmann, K. (2016). *Religious NGOs in international relations: The construction of 'the religious' and 'the secular'*. London: Routledge.

Marshall, K. (2013). *Global Institutions of religion: Ancient movers, modern shakers*. New York, NY: Routledge.

McGrath, M. (2019, July 24). Climate change: 12 years to save the planet? Make that 18 months. *BBC News*. Retrieved from https://www.bbc.com/news/science-environment-48964736

Meditation Room. (2019). United Nations Gifts. Retrieved from https://www.un.org/ungifts/content/meditation-room

PBS News Hour. (2015). Full text of Pope Francis' speech to United Nations Sep 25. Retrieved from https://www.pbs.org/newshour/world/full-text-pope-francis-speech-united-nations

Quaker United Nations Office. (2019, March). A government official's toolkit: Inspiring urgent climate action. Retrieved from https://quno.org/resource/2019/3/government-officials-toolkit-updated-a5-edition

Religion Communicators Council. (2019, August). United Nations discovers that faith-based organizations also support sustainable development goals. Retrieved from https://www.religioncommunicators.org/united-nations-discovers-that-faith-based-organizations-also-support-sustainable-development-goals

Sachs, J. (2019, August 23). *Moderator's remarks: Plenary IV: Advancing shared well-being by promoting integral human development and protecting the Earth*. Lindau: Religions for Peace 10th World Assembly.

Spencer, A. (2016, April 18). *Confluence of the Baker River and the Neff river, Aysén, Patagonia*. YouTube. Retrieved from https://www.youtube.com/watch?v=N5i01cLOeHo

Stensvold, A. (2016). *Religion, state and the United nations: Value politics*. London: Routledge.

Terra Galleria. (2019). Fresco memorializing the founding of the United Nations, Grace Cathedral San Francisco, California, USA. Retrieved from https://www.terragalleria.com/california/picture.usca43266.html

The United Nations System. (2019, July). United Nations department of global communications. Retrieved from https://www.google.com/url?sa=t&rct=j&q=&esrc=s&source=web&cd=1&ved=2ahUKEwjgprCIt_3mAhUSw1kKHbyYAcEQFjAAegQIARAC&url=https%3A%2F%2Fwww.un.org%2Fen%2Fpdfs%2F18-00159e_un_system_chart_17x11_4c_en_web.pdf&usg=AOvVaw36ZKOOrf3vasTCRxg4VlQG

The Universal Declaration of Human Rights. (2019). The Eleanor Roosevelt papers project. George Washington University. Retrieved from https://www2.gwu.edu/~erpapers/humanrights/udhr/.

Thunberg, Greta [@GretaThunberg]. (2019, August). *The @UN has sent out one boat for each of the 17 sustainable development goal to greet us! Thank you!* [Tweet; photo]. Twitter. Retrieved from https://twitter.com/GretaThunberg/status/1166791436538449922

UN Multimedia. (2019). Millennium world peace summit of religious, spiritual leaders meets at headquarters. Retrieved from https://www.unmultimedia.org/photo/detail.jsp?id=304/304300&key=274&query=summit&so=0&sf=date

United Nations. (1992). UN Framework Convention on Climate Change (full text). Retrieved from http://unfccc.int/files/essential_background/background_publications_html pdf/application/pdf/conveng.pdf

United Nations. (1993). Convention on Biological Diversity (full text). Retrieved from https://www.cbd.int/doc/legal/cbd-en.pdf

United Nations. (1994). Convention to Combat Diversification (full text). Retrieved from https://www.unccd.int/sites/default/files/relevant-links/2017-01/UNCCD_Convention_ENG_0.pdf

United Nations. (1997). Kyoto Protocol (full text). Retrieved from http://unfccc.int/resource/docs/convkp/kpeng.pdf

United Nations. (2015). Paris Agreement (full text). Retrieved from https://unfccc.int/files/meetings/paris_nov_2015/application/pdf/paris_agreement_english_.pdf

United Nations. (2018). Global Compact on Migration (full text). Retrieved from https://refugeesmigrants.un.org/sites/default/files/180711_final_draft_0.pdf

United Nations. (2019). UN Charter (full text). Retrieved from https://www.un.org/en/sections/un-charter/un-charter-full-text/

United Nations. About the sustainable development goals (2019). Retrieved from https://www.un.org/sustainable development/sustainable-development-goals/

United Nations. (2019, August). Overview. Retrieved from https://www.un.org/en/sections/about-un/overview/index.html

United Nations. Climate Change (2019). Retrieved from https://www.un.org/en/sections/issues-depth/climate-change/. Accessed on August 13, 2019.

United Nations Environment Programme. (2018). List of Environmental Faith-Based Organizations and Resources. Retrieved from https://wedocs.unep.org/handle/20.500.11822/25987. Accessed on August 1, 2019.

United Nations Environment Programme. (2019a). Faith for Earth initiative. Retrieved from https://www.unenvironment.org/about-un-environment/faith-earth-initiative

United Nations Environment Programme. (2019b). SDG 13 on Climate Action. Retrieved 8/2019 from http://wedocs.unep.org/bitstream/handle/20.500.11822/22748/13_Climate%20ActionFINAL.pdf?sequence=1&isAllowed=y

United Nations Environment Programme. (2019c). The Annual Kofi Annan Faith Briefing Agenda 2019 15 July

2019 New York, United States of America. Retrieved from https://www.unenvironment.org/events/conference/annual-kofi-annan-faith-briefing-2019

United Nations Environment Programme. (2019d). UN Interagency Task Force on Religion and Development - Annual Report 2018. Retrieved from https://www.unenvironment.org/resources/annual-report/un-interagency-task-force-religion-and-development-annual-report-2018

United Nations Secretary General Statements and Messages. (2009, October 26). Faith-based organizations can help deliver United Nations message for renewed cooperation to tackle climate change, other crises, Secretary-general says. Retrieved from https://www.un.org/press/en/2009/sgsm12567.doc.htm.Accessed on August 1, 2019.

United Nations Youtube Channel. (2018, September 10). Secretary-General António Guterres calls for global action on climate change. Retrieved from https://www.youtube.com/watch?time_continue=12&v=VNe-jBVij-g&feature=emb_logo

United States Conference on Catholic Bishops. (2019b). Laudato Si: On Care for our Common Home Retrieved from http://www.usccb.org/about/leadership/holy-see/francis/pope-francis-encyclical-laudato-si-on-environment.cfm. Accessed on August 9, 2019.

Weiner, M. (2010, September). Religion and the United nations – introduction. *CrossCurrents*, 292–296.

Whitney, L. & E. (2012, June). Faith based Statements on climate change, June 2012. Citizens Climate Lobby. Retrieved from www.citizensclimatelobby.org. Accessed on August 1, 2019.

12

DIFFERENT VOICES: ONE CALL

Sally Myers

I had the privilege of attending the 2019 Moana: Water of Life conference, and the pleasure of meeting most of the contributors to this volume. Consequently, when I read their chapters I can hear their voices in my imagination, conjure up their faces and remember their radiance. The 2019 conference was very much a conversation, earthed in human experiences and permeated with the personalities of real people. This is impossible to capture completely in a book, but we have attempted it insofar as possible by asking people to write 'as themselves', and we have deliberately left unveiled the different patterns of thought and underlying frameworks of understanding, which might have been lost had we applied a more flattening editing style. In this chapter I explore the overarching approaches of the contributors to the phenomena of, and dialogue, about climate crisis. (This is limited to the specific contributions here, and there are of course differences in the worldviews and methodological approaches of other Pacific Island State (PIS) researchers, environmental scientists, educators and theologians.) I briefly consider how these differences might be barriers to mutual understanding. However, what is striking when analysing the perspectives of the seemingly disparate contributors is not what separates them but what unites them. In fact, the whole process has been something of an ecological lesson, as will be seen.

In the first two chapters, indigenous scholars from the Pacific Island Region (PIR) demonstrate how crucial the Christian faith is to them, and to meaning-making and decision-making in the area. God is prioritised above all else; viewed as being omnipotent and immanent, and especially present within Moana, the ocean. The whole of physical life is seen in relation to this spiritual reality, God's promises and the notion of eternity. Scripture is read both for comfort and confirmation of God's will. This does not, however, result in climate denial. The contributors are realistic about the climate crisis. However, they are not despondent. On the contrary, it is their strong faith that provides them with hope and motivates resilience and adaptation to their immediate changing circumstances, rather than moving away.

In contrast, the chapters written by UK environmental scientists are focused on physical reality. Their contributions are founded upon empirical data, which has been carefully evaluated against agreed measurements and repeatedly checked and verified. They offer predictions and models based upon scientific observations of previous patterns, rather than faith in the future. Indeed, the conclusions they draw are not easy reading, nor is it fair to say, particularly hopeful. Although the call for mitigating action is loud and clear, there is acknowledgement that the message from science is frightening, hard to assimilate and overwhelming, but nevertheless extremely urgent.

The chapters by educators focus on the importance of informing people about, and equipping them to deal with, climate change. Understandably then, they speak of the need for much greater environmental literacy. However, it is the teaching methodology they prioritise that is most notable. Both chapters focus on the co-construction of knowledge between educator and learner, and through this, empowerment of response through authority and agency. Although the content of the education is viewed as vitally important, so too is inspiring and preparing the students to take the necessary actions to respond to climate change. In both chapters, social and environmental justice, especially that centred around care for the vulnerable, is prioritised.

The two theological chapters that follow are again written by practising Christians. The first seeks to personally engage the reader with reflective exercises, anecdotes and an invitation to individual and corporate relationship with scripture. As with the first two chapters, climate crisis is presented as a spiritual crisis. Again, the emphasis is on right relationship with God; this time urging human beings to stop playing God, repent and work to restore the damage that has been caused by self-assertion and consequent exploitation of the earth and other humans. The second theological chapter in contrast argues that the climate crisis is a religious crisis, caused by an unbalanced reading of scripture, which has in turn led to church doctrine that has proved damaging to the environment. It offers an alternative reading of the Bible that challenges the traditional religious view of human beings as set over and above the rest of creation, and which resets the relationship to one of mutuality and common care.

The final three chapters are less about theory and cause, and more about solutions. They provide oversights of the role of technology, both physical and virtual, and politics in the form of the United Nations, and then offer practical ways in which individuals and organisations might begin to address the climate crisis. Whereas the scientists and theologians take the long view and set the current difficulties in the context of previous millennia, the educators and practitioners take a more sleeves-up-let's-sort-this-out attitude.

There are then very clear differences in approach. Cognitive frameworks and understandings of agency are centred in different loci: science in the physical realm; faith in the spiritual realm; and human initiative in the classroom, at home and in world politics. Different methodologies are employed: observation and measurement, reflection, scriptural reasoning, constructionism, theoretical analysis and information-gathering and consensus-seeking through networking. Sweeping overviews are presented alongside individual responses. Global information sits next to intimate personal insight. In addition, the contributors to the book are of different ages and the products of different continents, socio-economic contexts and cultural histories. Add into the mix a diverse range of characters and personalities, and it is amazing that we found any common ground at all; but we did. The urgency of the problem and the meeting together over food as fellow human beings will have helped. However, there are also some underlying patterns and themes that cut across the differences in approach and background.

For good or ill, all of the chapters talk about connectedness. This includes recognising: that faith is not compartmentalised, but infuses daily lives and decision-making; the interrelationship within nature, especially explored here with reference to changing patterns in rivers and oceans; the interconnection between the physical and spiritual realms; physical and virtual connection between individuals, countries and organisations through technology; the symbiotic relationship between humans and the environment. Furthermore, it was also recognised, again across the disciplines, that there is often a delicate balance involved in these interconnections, which is vulnerable to small changes that may set in motion much larger consequences. Any sustainable solution or adaptation to the climate crisis is reliant upon a deep understanding of the connectedness of the earth's functioning and a connected response. Currently both problem and solution are subject to disconnection.

Human beings naturally separate and categorise in order to make sense of their world. However, this is often augmented by postulating hierarchy. This sometimes results in not only disconnecting phenomena but also privileging one above another. In relation to our subject matter this might include: humanity over the rest of creation, science over humanities, religious authorities over followers, teachers over students, fact over feeling, reason over faith, expert over citizen and economy over justice and welfare. Indeed, it might be argued that it is an inappropriate top-down use of science and religion that has led to the problems we now face. Interestingly, all of the contributors look in some way to rebalance these relationships.

Across this volume there is a theme of inverting established hierarchies, recognising that: it is the people on the ground who really know the situation and can provide and deliver the solutions to it; the spiritual realm holds the key to putting right the difficulties caused by abuse of power in the physical realm; educators look to their students to help them create knowledge; creation is to be honoured rather than exploited and seen as revelation of God and the way to a good life.

There also appears to be a rebalancing of internal and external realities. The physical manifestations of the climate crisis undoubtedly need to be urgently addressed. However, in order for any changes to be real and sustainable, the internal perspectives and motivations of people need to change too. Whether religious language is used or not there is a clear message that we need to take seriously our personal resonance with the earth, and rebalance our relationship with it, not least in a reordering of consumption patterns. It is one thing to be able to dissect something and see the mechanics of how it works. It is another to take time to apprehend its spirit and to know it. Rather than being an object to be exploited, the earth might be seen as a living being, engendering an attitude of gratitude rather than entitlement among those of us who share its life.

In times of urgency and fear the tendency is to resort to binary and uncompromising thinking and action. This narrowing of vision is sometimes necessary. However, it would seem that developing a deeper understanding and respect for diverse meaning-making systems will not only help us listen to and understand each other more fully, but it may also be the key to enabling the coordination and collaboration necessary to if not resolve, then at least adapt to the climate crisis. Unsurprisingly, as with all creation, although the disciplines represented by the contributors to this volume may appear at first glance to be a rather random collection of discrete entities, if we work together and with others, we may yet find an eco-solution to eco-dysfunction.

CONCLUSION

It might reasonably be argued that faith and science have contributed greatly to climate change and therefore have no small responsibility for addressing the issue. In this volume, science and faith, alongside educators and practitioners, have sought to find ways of communicating and working with one another to do just that. We have seen how the people of the Pacific Island Region have modelled the cooperation needed. They have been working together under threat and uncertainty about rising sea levels for a sustained period and are already engaged with resilience and adaptation 'on the ground'. We hope that this volume has provided an introduction to the different perspectives on and potential for this collaboration on a wider scale.

There is much missing from this volume; subjects for future consideration might include the role of art in raising awareness and combating climate stress; research with those who hold the earth in reverence across different faiths and traditions; further exploration of the place of creation in scripture and Christian tradition; consideration of environmental ethical reasoning across the disciples; environmental curricula development. In the meantime, we would like to commend continued consideration of the importance of story and human contact; the interconnection of all life, seen and unseen; the need for nurture of our inner worlds as well as the outer world we share; the need to attend to necessary balance within creation and our part in it.

The role of church and education in adaptation to and mitigation of climate change remains critical. We applaud the efforts already being made and urge the following actions: enhanced eco-responsible use of money, buildings and land; building empowerment into education; tackling fear and inertia though prayer, preaching, research and celebration; shaking hands and showing up in support of community, national and international projects; and, not least, the communication of the possibility of transformation through hope and united action.

As we have read, small changes set in motion much larger consequences, and if we embrace this natural way of things, we may yet be able to reverse the situation.

AFTERWORD

MOANA: WATER OF LIFE

Come and See: An Invitation to Life-giving Engagement

The words 'Come and see' are words of Jesus to his disciples in John's Gospel (John 1.30). They are an invitation to life-giving engagement. They are an invitation to experience a new kind of living and being and to set out on an adventure. As I look back, I recall with gratitude the work of the United Society Partners in the Gospel (USPG) and its worldwide mission; I look back and recall the response of the Council of the USPG when I extended an invitation to the Council (comprising archbishops, bishops and other leaders in the Anglican Communion) to visit Fiji for the first time. There was deep concern in the Council over global warming and the impact of climate change on the environment and the lives of so many people. My invitation was for the Council to meet in the Diocese where the effects of climate change in terms of intensity of cyclones and the threat of rising sea level were a growing reality. The Diocese of Polynesia embraces some of the most vulnerable nations to climate change. The USPG Council responded to an invitation to come and see.

The Rt Rev Christopher Lowson was a member of the USPG Council, and he came to Fiji as representative of the Archbishop of Canterbury. Bishop Chris not only came for the first time to Fiji for the USPG Council, he returned to explore the establishment of a partnership between the Diocese of Lincoln and the Diocese of Polynesia. At that time he was putting forward the idea of an International Conference to address climate change. Bishop Chris travelled widely sharing our experience in Fiji and Tonga. He travelled by air and by boat. He travelled along bumpy roads into the interior of Fiji and to the coastal regions. He sat with our people. He listened. During his time and experience in the Diocese the seeds for the International Conference began to grow in his heart and mind. Bishop Chris had already a vision for an International Conference on Climate Change. This vision was strengthened as he saw that the Diocese of Polynesia could also contribute and so the Conference which combined the efforts of the Diocese of Lincoln, the universities in Lincoln and the Diocese of Polynesia was conceived. In Polynesia, we give thanks for Bishop Chris' vision and passionate commitment.

An invitation from the Principal of Westcott House to my wife, Reverend Sue Halapua, to be Chaplain at the Westcott House facilitated a time for both of us to be in England. This enabled me to be part of the Committee planning the Conference. It was proposed that a delegation from Polynesia to International Conference included members from the wider Province of Aotearoa, New Zealand, and Polynesia. The delegation was to include people from the different ethnic groups and from different situations. The delegation was to include laity and clergy, students of St John' Colleges in Suva, Fiji, and Auckland, New Zealand, and academics. There was a large group of young people, some of whom are actively involved in providing help in time of tropical cyclones. As Polynesians, we were intended to demonstrate working together to address the serious issues of climate change and committed to working together with Lincoln, the Diocese and the universities.

We were grateful for the welcome expressed by the Rt Rev Dr David Court, Acting Bishop of Lincoln. We were grateful for the huge hospitality of Lincoln.

I introduced the conference and the delegates to one another. I explained how Moana: Water of Life, the Polynesian word for 'ocean', is hugely significant for the peoples and environment of the Pacific. The five oceans of the world flow constantly giving life. They are interconnected and speak of our interconnectedness as people of this Blue Planet Earth.

My introduction was not an isolated contribution but formed part of the Polynesian contributions, which included a Biblical Paper by Dr Emily Colgan. Our young people presented a dynamic way of working together to meet the needs of people in time of tropical cyclones. Our young people, towards the end of the conference, demonstrated their faith and leadership as they participated in worship at the Thanksgiving Service of the Cathedral. They used their voices and the movements of traditional Polynesian dance to pray the Lord's Prayer. Movingly and indicating ongoing leadership, the new Archbishop, the Most Rev Fereimi Cama, led them in the Prayer.

I am deeply aware of the efforts of Pacific people and others to achieve an agreement in Paris COP21. A significant number of people attending the Conference, including Professor Elisabeth Holland, a key member of the Intergovernmental Panel on Climate Change, had worked hard and together behind the scenes to influence and to achieve an agreement to reduce emissions to meet global temperature targets among the nations gathered in Paris.

Moana: Water of Life called for deep listening and a new intentional togetherness across disciplines and diversity as we navigate into the future and tackle huge challenges. The listening across the disciplines of science and theology, the listening to the perspectives and sharing of the experiences of different peoples are put forward in order to enable new visions for our Planet Earth and moving forward actively in solidarity. We are called not to be locked in narrow ways of thinking or limited perspectives as we address with courage the great challenges of our time. We are to be open to larger horizons.

This publication has recorded some of the contributions to meeting that challenge. However, there is much more still to be done. What is clear is the continuing need for deep listening, to the climate scientists, to the educators, to the youth and perhaps most of all to Moana. We need to take seriously and to respect the God within nature, to hear the pain that is generated by human misuse and to redress this. However, there is more potential benefit than simply halting the difficulties caused by global warming. If we listen deeply, we can perhaps hear and learn some deep lessons for what it means to live a good life, in balance with creation and with God's will, however defined. My hope is that this book will provide the launch pad for these discussions.

Winston Halapua
(Retired Archbishop of Polynesia)

INDEX